Dimensional Analysis and Similarity in Fluid Mechanics

Series Editor
Roger Prud'homme

Dimensional Analysis and Similarity in Fluid Mechanics

Nord-Eddine Sad Chemloul

iSTE

WILEY

First published 2020 in Great Britain and the United States by ISTE Ltd and John Wiley & Sons, Inc.

ISTE Ltd
27-37 St George's Road
London SW19 4EU
UK

www.iste.co.uk

John Wiley & Sons, Inc.
111 River Street
Hoboken, NJ 07030
USA

www.wiley.com

Library of Congress Control Number: 2020940665

British Library Cataloguing-in-Publication Data
A CIP record for this book is available from the British Library
ISBN 978-1-78630-596-1

Contents

Foreword

It has been my great pleasure to read this book, a series of exercises and a course review, written by Nord-Eddine Sad Chemloul.

The book addresses dimensional analysis and similarity, tools for solving specific problems in fluid mechanics, and is intended for students of mechanical engineering, civil engineering, maritime engineering, and other similar fields.

The book's author has taught this subject for over 25 years and has acquired a great deal of experience, allowing him to perfect his course over the years.

The book is firmly structured and well written, making it an easy read. It is divided into three chapters. The first chapter concerns the calculation of dimensions, the verification of the homogeneity of relationships, and the conversion of units. The second chapter introduces the homogeneity method (or Rayleigh method) and the Vaschy-Buckingham method (also known as the π method), which are used in dimensional analysis. The third chapter provides a study of the comparison of fluid flows using similarity.

The book's numerous exercises, the product of the numerous works and tests prepared by author, are accompanied by detailed and thoroughly commented corrections. This book is also intended for educators.

I highly recommend it for students, since it provides them with the essential elements of dimensional analysis and similarity as applied to fluid mechanics.

Driss NEHARI
Professor at the University Center
of Ain Temouchent, Algeria

Preface

The theories of fluid mechanics can at times be quite complicated. To better understand them, it is necessary to use exercises to apply them.

The three chapters that make up this book begin with a simple review of the essential knowledge, supplemented by examples.

Fifty exercises are worked out and include commentary. They cover almost all areas of fluid mechanics: flow through orifices, dynamic boundary layers, the behaviors of fluids on solid faces, etc.

These exercises range from the simplest, on the more pedagogical side, to the most difficult, on the practical side. They are intended for all students seeking to familiarize themselves with the discipline of fluid mechanics, in the field of technological sciences, science, and physics.

We would be happy to hear any critiques and suggestions that may be made regarding the level, presentation, and choice of exercises given in this book. We would like to clarify that the choice of exercises has been made with the hope of achieving a compromise between the general education and training of students and the practical implementation of the knowledge they obtain.

Nord-Eddine SAD CHEMLOUL
July 2020

Introduction

The analytical methods (or models) that are used in solving the problems of fluid mechanics are not always satisfactory (as they do not give good results) in practice. This is because it is necessary to use simplifications or detailed and onerous analyses.

The alternative to this is to call upon past experience and deduce correlations applicable to all cases of the same type of problem. In many cases, the experimental conditions in the laboratory are not the same as those in real life (for example, the dimensions of the centrifugal pump model and its prototype, or the different fluids), and in these cases, dimensional analysis is used. This analysis allows for dimensionless correlations of dimensionless numbers to be obtained, which are generally applicable to all practical cases. In other words, these dimensionless correlations can be applied under dynamic conditions, similar to those in which they were established by using, for instance, a different fluid.

Dimensional analysis is a practical method for verifying homogeneity (the method devised by John William Strutt Rayleigh) of a physical formula through its dimensional equations. This is achieved by subdividing the physical quantities included in the formula into a product of fundamental or basic values: length, duration, mass, electrical intensity, etc., each of which are irreducible with respect to all others.

Dimensional analysis is based on the principle that only the values of the same dimension can be compared or added; a length may be added to another length, but it cannot be said to be greater than, or less than, a measurement of mass. A physical law cannot change, except in the numerical values of its constants when expressed in other units. This is mathematically demonstrated by the theorem of Aimé Vaschy and Edgar Buckingham.

In basic physics, dimensional analysis makes it possible to obtain the form of an equation from assumptions about the quantities that govern the state of a physical system. Then, the hypotheses can be validated by a complete theory. For applied sciences, dimensional analysis is the basis for modeling through replicas and the study of effects of scale.

Similarity in fluid mechanics plays a very important role in the design and construction of large-scale and high-cost installations, and structures in many fields such as aeronautics, hydraulics, chemical engineering, and others. Thus, it is important to have the necessary tools to accurately define all characteristics of the work that would allow us to achieve the desired outcomes.

Therefore, we must attempt to represent the true operating conditions of the work using a model. The accuracy of the results obtained will depend on the level of quality of this representation. However, the problems addressed in fluid mechanics show that it is very difficult to correctly translate the characteristics of a given flow at any time and at any point, especially when this flow is turbulent. This means it is impossible to correctly write about and address the basic equations, because the model is only an approximate representation of the actual functionality. In this sense, two possibilities have been provided for: solving simplified equations mathematically, or the conducting of a laboratory study of a concrete representation of the work. These two methods are in fact complementary, and it is always in our best interest to use them in conjunction with each other.

In this book, only the second method is used. This method involves creating a physical model of the proposed installation or structure, which defines the similarity. The dimensions of the model are generally smaller than those of the work, meaning that the model is reduced.

1

Homogeneity of Relationships and Conversion of Units

1.1. Introduction

The creation of a system of units requires the definition of basic units, their values, and the units derived from them. In mechanics, the units used are length, mass and time, but other options are possible as well, such as length and time, force and time, or mass, speed and time.

The first general conference on weights and measures was held in 1889 at the headquarters of the BPIM (*Bureau international des poids et mesures* or International Office of Weights and Measures), at the Breteuil pavilion in Sèvres (in the suburbs of Paris). In this conference, new international prototypes of the meter and kilogram were officially adopted and filed with the Office.

In 1960, the 11th General Conference of Weights and Measures established the International System (SI), in which the rules for prefixes, derived units and other indications were established. The SI is based on a choice of seven well-defined base units that the convention considered dimensionally independent: the meter, the kilogram, the second, the ampere, the kelvin, the mole and the candela. Derived units are formed by combining the base units according to the algebraic relationships connecting the corresponding quantities. The names and symbols of some of these units may be replaced by special names and symbols, which may be used to express the names and symbols of other derived units.

In November 2018, the International System of units later underwent a significant revision at the 26th General Conference of weights and measures, the culmination of nearly 250 years of consideration given to finding the best way to define a system of units of measurement that would best reflect the natural world.

The General Conference of weights and measures, established at the end of the 19th Century, meets every four to six years to discuss, and possibly modify, the SI, standardizing the units on a global scale. The new changes were applied as of May 20, 2019.

Quantity	Name	Symbol	Dimension
Length	Meter	m	L
Mass	Kilogram	kg	M
Time	Second	s	T
Intensity of electric current	Ampere	A	I
Thermodynamic temperature	Kelvin	K	Q
Quantity of matter	Mole	mol	N
Intensity of light	Candela	cd	J

Table 1.1. *Basic SI units*

1.2. Definitions of the basic SI units

1.2.1. *Definition of the meter as adopted in 1983*

The “meter” is the “length” of the distance light travels in a vacuum for a period of 1/299792458ths of a second. From this, it can be determined that the speed of light in a vacuum is exactly: $c_0 = 299{,}792{,}458$ m/s.

1.2.2. *Definition of the kilogram*

The “kilogram” is the unit of mass. It is equal to the mass of the international prototype of the kilogram. The term “weight” refers to a quantity that is of the same nature as a force; the weight of a body is the product of the mass of that body multiplied by the acceleration due to gravity. Specifically, the normal weight of a body is the product of the mass of this body multiplied by the normal acceleration from gravity. The number adopted by the International Service of Weights and Measures for the value of normal acceleration from gravity is 980.665 m/s^2.

The kilogram is currently defined as the mass of an Iridium platinum cylinder (90% platinum and 10% iridium) 39 mm in diameter and 39 mm high declared the SI unit of mass in 1889 by the BIPM. This unit of measurement is the last SI unit to be defined using a man-made physical standard. It is stored under three sealed glass bells, and it is only removed from this covering for calibrations (an operation that has taken place only three times since its creation). As of May 20, 2019, the kilogram has been defined on the basis of Planck's constant (h)[1] from quantum physics, measured on the Kibble[2,3] scale at $6.626069934 \times 10^{-34}$ $kg.m^2.s^{-1}$.

1.2.3. *Definition of the second adopted in 1967*

The "second" has the duration of 9,192,631,770 periods of radiation from the transition between the two hyperfine levels of the base state of a cesium 133 atom. As a result, the frequency of the hyperfine transition of the base state of the cesium atom is equal to 9,192,631,770 hertz (Hz). Thus, we obtain exactly ν (hfs[4] Cs) = 9,192,631,770 Hz. At its 1997 session, the International Committee on Weights and Measures confirmed that this definition refers to a cesium atom at rest, at a temperature of 0 K (Kelvin[4]).

---

1 Max Planck, born Max Karl Ernst Ludwig Plank, on April 23, 1858, in Kiel, in the Duchy of Schleswig (Germany), living until October 4, 1947, was a physicist. Planck's constant h plays a central role in quantum mechanics: it links the energy of a photon to a photon's frequency.

2 The scale created by Bryan Peter Kibble is a device that allows for the conversion, with nine significant figures, of mechanical power into electrical power and vice versa. Kibble, born on August 10, 1938, in Letcombe Regis (England), who lived until April 28, 2016, was a physicist and an expert in metrology

3 Hyperfine splitting of cesium.

4 William Thomson, better known as Lord Kelvin, born on June 26, 1824, in Belfast (Northern Ireland), who lived until December 17, 1907, was a physicist best known for his works on thermodynamics.

1.2.4. *Definition of the ampere adopted in 1948*

The "ampere"[5] is the intensity of a constant current that is produced between two parallel conductors following straight lines of infinite length and of a negligible circular cross-section placed at a distance of 1 meter from each other in a vacuum between these conductors, a force equal to 2.10^{-7} newtons[6] per meter of length. As a result, the magnetic constant, also known as vacuum permeability, is equal to exactly $4\pi.10^{-7}$ Henrys[7] per meter ($4\pi.10^{-7}$ H/m). The Ampere is linked to the elementary charge (e), the electric charge of a proton. The mole, the unit for quantities of matter, used mainly in chemistry, is defined directly through the determining of Avogadro's number (AN)[8].

1.2.5. *Definition of Kelvin adopted in 1967*

The Kelvin, a unit of thermodynamic temperature, is the fraction (1/273.16) of the thermodynamic temperature of the triple point of water. As a result, the thermodynamic temperature of the triple point of water is exactly 273.16 Kelvin (273.16 K). Other changes that were made: the Kelvin scale, measured based on water, was redefined based on the Boltzmann[9] constant (k), related to the measurement of the thermal agitation of the fundamental constituents of a body.

---

5 André-Marie Ampère, born in Lyon (France) on January 20, 1775, and who lived until June 10, 1836, was a mathematician, physicist, chemist and philosopher. He contributed to the development of mathematics by introducing it into physics. He made important discoveries in the field of electromagnetism.

6 Isaac Newton, born on January 4, 1643, in Woolsthorpe (England), and who lived until March 31, 1727, was a philosopher, mathematician, physicist, alchemist, astronomer and theologian. His book, *Philosophiæ Naturalis Principia Mathematica*, published in 1687, is considered one of the greatest works in the history of science.

7 Joseph Henry, born on December 17, 1797, in Albany (New York) and who lived until May 13, 1878, was a physicist who discovered self-induction and the principle of electromagnetic induction from induced currents.

8 Lorenzo Romano Amedeo Carlo Avogadro, Count of Quaregna and Cerreto, known as Amedeo Avogadro, was a physician and chemist, born in Turin (Italy) on August 9, 1776, and who lived until July 9, 1856.

9 Ludwig Eduard Boltzmann, born on February 20, 1844, in Vienna (Austria) and who lived until September 5, 1906, was a physician and philosopher, considered to be the father of statistical physics, and an avid proponent of the existence of atoms.

1.2.6. *Definition of a mole*

The mole[10] is the amount of matter in a system containing as many indivisible entities as there are atoms in 0.012 kilograms of carbon 12; the symbol for this unit is “mol”. When using the mole, the individual entities must be specified, and may be atoms, molecules, ions, electrons, other particles, or specified groups of such particles. In this definition, it is understood that this refers to unlinked carbon 12 atoms, at rest, and in their base state. As a result, the molar mass of carbon 12 is equal to 0.012 kilograms per mole (0.012 g/mol).

1.2.7. *Definition of the candela adopted in 1979*

The Candela[11] measures the luminous intensity in a given direction of a source that emits energy in monochromatic rays in that direction; it is 1/6, 831ths of a watt[12] per steradian[13]. From this, it can be determined that the spectral luminous efficiency of monochromatic rays of a frequency of 540×10^{12} hertz is equal to 683 lumens[14] per watt (683 lm/W = 683 Sr/W).

1.3. Additional quantities and SI derived quantities

Table 1.2 gives the two additional quantities that have been introduced to ensure the coherence of the system.

Table 1.3 shows the derived quantities in the International System generally used in fluid mechanics and heat transfer.

---

10 The name of the unit of quantities of matter, the mole, originates from the abbreviation of the German word *Molekül*: “Mole” (with a capital “m”, and followed by a rapidly abandoned point) was proposed in 1893 by chemist Wilhem Ostwald to indicate “the weight in grams numerically equal to the molecular weight of a given substance”. The term “Mole” first appeared in 1902 as the English equivalent of the German term.

11 The *candela* (abbreviated as “cd”, from the Latin word meaning “candle”) is one of the seven basic SI units.

12 James Watt, born on January 19, 1736, in Greenock (Scotland) and who lived until August 25, 1819, in Heathfield Hall (in his house in Handsworth, England), was an engineer whose improvements to the steam engine were one of the key advances in the industrial revolution.

13 The Steradian (symbol: “sr”) is the SI derived unit for the measurement of solid angles.

14 The *lumen* (from the Latin word for “light”) is the SI derived unit for luminous flux.

Quantity	Name	Symbol	Dimension
Angle of a plane	Radian	Rad	–
Solid angle	Steradian	sr	(Ω)

Table 1.2. *SI derived quantities*

Quantities	Unit	Name	Dimension
Speed	m/s	Meter per second	LT^{-1}
Angular velocity	rd/s	Radian per second	T^{-1}
Acceleration	m/s^2	Meter per second squared (meter per second per second)	LT^{-2}
Force	N	Newton	MLT^{-2}
Moment of force	N. m	Newton-meter	ML^2T^{-2}
Surface tension	N/m	Newton per meter	MT^{-2}
Work, energy	J	Joule	ML^2T^{-2}
Power	W	Watt	ML^2T^{-3}
Pressure	Pa	Pascal	$ML^{-1}T^{-2}$
Quantity of movement	N.s	Newton-second	MLT^{-1}
Dynamic viscosity	Pa.s	Pascal-second	$ML^{-1}T^{-2}$
Kinematic viscosity	m^2/s	Square meter per second	L^2T^{-1}
Mass	kg	Kilogram	M
Density	Kg/m^3	Kilogram per cubic meter	ML^{-3}
Volumetric flow	m^3/s	Cubic meter per second	L^3T^{-1}
Mass flow rate	Kg/s	Kilogram per second	ML^{-1}
Time	s	Second	T
Surface area	m^2	Square meter	L^2
Volume	m^3	Cubic meter	L^3
Temperature	K	Kelvin	Q
Thermal capacity	J/K	Joule per kelvin	$L^2T^2 Q^{-1}$
Thermal conductivity	W/m.K	Watt per meter-kelvin	$MLT^3 Q^{-1}$
Thermal convection	W/m^2K	Watt per square meter-kelvin	$MT^{-3} Q^{-1}$

Table 1.3. *Some of the SI derived quantities used in fluid mechanics and heat transfer*

1.4. Rules for the use of units

1.4.1. *Unit name*

All names of units, even those derived from the names of noteworthy figures, are considered as common nouns: pascal, newton, hertz, volt[15], ampere, henry, weber[16], watt, joule, pascal[17], newton, hertz, coulomb[18], etc.

The initials are written in lowercase, and are given in plural if they apply to a number greater than or equal to 2. For example: 1.9 volt, 3 amps, 1.4 newton, 5 watts, 3 henrys, etc.

The names of the units derived from the names of notable scientific figures are masculine in gender (one joule, one Ampere, one henry).

No qualifiers should be added to a unit name (the term “linear meter” is not used).

When one quantity is the product of two others, and neither of them is a quotient, the name of the unit is obtained by joining the two corresponding units with a hyphen (making sure not to use a forward slash, which indicates a quotient). For example, electrical energy that is the product of power and time can be expressed as a “watt-hour”. In the case of very common units, the two names can be combined: a “watthour” or a watt-hour. The plural mark is to be added to the two component names in the case of a hyphen and to the last one in the case of contiguous nouns: watt-hours or watthours, meters-newtons.

When one quantity is the quotient of two others which are not quotients themselves, the name is obtained by placing the word “per” (and not the

15 The name “Volt” was created in tribute to Alessandro Volta, an Italian inventor who invented the voltaic pile (battery) in 1800.

16 Wilhelm Eduard Weber, born on October 24, 1804, in Wittemberg (Germany) and who lived until June 23, 1891, was a physicist who developed an original theory of electromagnetic interaction.

17 Blaise Pascal, born on June 19, 1623, in Clermont-Ferrand, and who lived until 1662, was a mathematician, physicist, inventor, philosopher, moral thinker and theologian.

18 Charles-Augustin Coulomb, born on June 14, 1736, in Angoulême, and who lived until August 23, 1806, was an officer, engineer and physicist who is the namesake for the unit for electric charge in SI.

symbol for division) between the units of the dividend and those of the divisor: kilometer per hour, meter per second.

1.4.2. *Unit symbols*

Following a number, a unit name can be replaced by its symbol: we may write 5 meters or 5 m, but it is necessary to write five meters (and not "five m").

A unit symbol should not be changed; in particular, it should never be given in plural: 15 kg (and not "15 kgs") and it should not be followed by a dot, except at the end of a sentence. A symbol must be placed after the numerical results in the case of decimal units: 26.3m (not "26m3"). This rule does not apply to units that are not decimal units: 12hrs, 15m, 30s.

1.4.3. *Compound symbols*

In the case of compound units, the symbol is represented by an algebraic expression in which each of the symbols plays the same role as the magnitude defining the equation; thus in the case of a product (vector or scalar) or quotient, the symbol is the product (vector or scalar) or the quotient of the symbols of the compound units. For example, the watt-hour is represented by the symbol (Wh) for the unit of electrical energy.

A velocity, the quotient of length and time, can be expressed with a unit whose symbol is m÷s (horizontal bar) or m/s (forward slash) or ($m.s^{-1}$).

Since the moment of a force is numerically equal to the vector product of the intensity of a force and a length, the symbol of the unit of moment can be written (m∧N or N.m). It should be noted that, according to ISO 31, the point between the two symbols should be at the midpoint of the height of the line of text: N·m.

The expression obtained can be transformed by applying the rules of algebra, with electrical resistance being defined by the expression (RS/l) where R is a resistance, S a surface and l a length, and the corresponding unit has the symbol ($W.m^2/m$) or (W.m) after simplification. Positive or negative exponents can also be used in these expressions: density is the quotient of a mass and a volume; the unit uses the symbol (kg/m^3 or $kg.m^{-3}$).

1.5. Exercises

1.5.1. *Exercise 1: calculation of dimensions*

1) Write the dimension of force.

2) Write the dimension of energy.

3) Write the dimension of a pressure (force per unit area).

4) Show that a measurement of pressure is one of volumetric energy. Can you find a formula that translates this result?

5) Write the dimension of the expression $(\rho V^2/2)$ where ρ is a density and V is the velocity of a fluid. Compare with the dimension of a pressure.

6) Write the dimension of ρgz with g as the acceleration from gravity z as the side. What do you conclude from this?

7) Can you give a physical formula that translates questions 4), 5) and 6)?

8) Check the homogeneity of the equation $V=\sqrt{2gh}$, where h is a height. What do you think the result expresses?

9) In the following relationship $\tau=\mu\dfrac{dV}{dy}$, τ is the shear stress, μ is the dynamic viscosity and $\dfrac{dV}{dy}$ is the speed gradient. Determine the dimension of the dynamic viscosity μ.

1.5.1.1. *Solutions*

1) The force is given by $F=m\gamma$, and its dimension is $[F]=[m][\gamma]$, with m as mass and γ as acceleration.

$$\begin{cases}[\gamma]=\left[\dfrac{\text{speed}}{\text{time}}\right]=\dfrac{LT^{-}}{T}=LT^{-2} \\ [\text{mass}]=M\end{cases} \Rightarrow [F]=MLT^{-2}$$

2) The dimension of a measure of energy E is:

$$E = \underbrace{F}_{\text{force}} \times \underbrace{l}_{\text{lenght}}$$

$$[E] = [F][l]$$

$$\begin{cases} [F] = MLT^{-2} \\ [l] = L \end{cases} \Rightarrow [F] = ML^2T^{-2}$$

3) The dimension of pressure P is:

$$P = \frac{\text{force}}{\text{unit area}} = \frac{F}{S}$$

$$\begin{cases} [F] = MLT^{-2} \\ [S] = L^2 \end{cases} \Rightarrow [P] = \frac{[F]}{[S]} = \frac{MLT^{-2}}{L^2} = ML^{-1}T^{-2}$$

4) the dimension of the relationship $\frac{\text{Energy}}{\text{volume}} = \frac{E}{v}$ is:

$$\left[\frac{E}{v}\right] = \frac{[E]}{[v]} = \frac{ML^2T^{-2}}{L^3} = ML^{-1}T^{-2}$$

COMMENT 1.1.– *The dimension of the relationship between energy and volume is that of a pressure. Pressure therefore represents a measure of energy per unit of volume, which represents one of the three forms of the Bernoulli equation expressed in joules per unit of volume. We recall that the Bernoulli equation expresses the conservation of total mechanical energy (kinetic energy + potential energy + pressure energy).*

For formulas translating the same result, we can cite the following: the manometric (or effective) pressure P_{man} *at a depth with respect to a free surface subjected to atmospheric pressure (taken as equal to zero) is* $P_{man} = \rho gh$.

With $[\rho] = ML^{-3}$, $[g] = LT^{-2}$ *and* $[h] = L$, *the dimension of the term* $P_{man} = \rho gh$ *is:*

$$[\rho gh] = ML^{-1}T^{-2}$$

The unit for this is:

$$\frac{kg}{ms^2} = \frac{kgm^2}{m^3 s^2} = kg\frac{m}{s^2}\frac{m}{m^2} = \frac{Nm}{m^3} = \frac{\text{Joule}}{\text{unit volume}}$$

We can also give the expression of the amount of work exchanged with a perfect gas.

5) The term $\rho\frac{V^2}{2}$ has the dimension.

We recall that the dimension of a constant is equal to the unit.

The dimension $\rho\frac{V^2}{2}$ of the relationship is that of a pressure, and is called dynamic pressure.

6) The dimension of the term ρgz, with z as the side.

$$\begin{cases} [\rho] = \dfrac{M}{L^3} \\ [g] = \dfrac{L}{T^2} \\ [z] = [\text{lenght}] \end{cases} \Rightarrow [\rho gz] = [\rho][g][z] = ML^{-1}T^{-2}$$

The dimension of the term ρgz is that of a pressure; it is referred to as potential energy of gravity or static pressure.

7) The relationship that translates questions 4), 5) and 6) is the total conservation of mechanical energy or Daniel Bernoulli's equation[19]:

$$P_{Total} = \underbrace{P_{static}}_{\substack{\text{energy from} \\ \text{pressure forces}}} + \underbrace{\rho g z}_{\substack{\text{potenial energy} \\ \text{of gravity}}} + \underbrace{\rho \frac{V^2}{2}}_{\substack{\text{kinetic energy or} \\ \text{dynamic pressure}}} = \text{constant}$$

This relationship represents the total pressure P_{Total}.

8) The homogeneity of the equation is $V = \sqrt{2gh}$.

The dimension of speed is $[V] = LT^{-1}$.

The dimension is $\sqrt{2gh}$.

$$\left[\sqrt{2gh}\right] = [2gh]^{1/2} = \underbrace{[2]^{1/2}}_{=1} [g]^{1/2} [h]^{1/2} = (LT^{-2})^{1/2} (L)^{1/2} = LT^{-1}$$

The equation $V = \sqrt{2gh}$ is therefore dimensionally homogeneous.

9) The dimension of dynamic viscosity is:

$$[\mu] = \left[\frac{\tau}{dV/dy}\right] = \frac{[\tau]}{[dV/dy]} = \frac{[\tau]}{[dV]/[dy]}$$

$$[\tau] = \frac{[\text{tangential force}]}{[\text{unit area}]} = \frac{MLT^{-2}}{L^2} = ML^{-1}T^{-1}$$

$$[dV] = [\text{speed}] = LT^{-1}$$

$$[dy] = [\text{length}] = L$$

19 Daniel Bernoulli, born in Groningen (Netherlands) on February 8, 1700, and who lived until March 17, 1782, was a doctor, physicist and mathematician. He formulated the fundamental theorem of fluid mechanics which bears his name, Bernoulli's theorem, in his book *Hydrodynamica* (1738).

And therefore:

$$[\mu] = ML^{-1}T^{-1}$$

COMMENT 1.2.– *From questions 5), 6) and 7), assuming that the side is zero, we obtain the relationship giving the dynamic pressure*:

$$\rho\frac{V^2}{2} = P_{Total} - P_{static}$$

which determines the velocity of the flow, or:

$$V = \sqrt{2\frac{P_{Total} - P_{static}}{\rho}}$$

In practice, the total pressure and the static pressure are measured using a Pitot tube[20]*, or separately using piezometric tubes.*

In question 8), the relationship $V = \sqrt{2gh}$ *is called Torricelli relationship*[21]*, and is obtained by applying the Bernoulli equation (with respect to a horizontal reference plane HRP) between the point (1) belonging to the free surface of a liquid contained in a reservoir and the point (2) belonging to an orifice located at a given depth below the free surface (see Figure 1.1), which is*:

$$\underbrace{\frac{V_1^2}{2g}}_{\substack{=\,0\text{, since the reservoir}\\ \text{has large dimensions}}} + \underbrace{\frac{P_1}{\rho g}}_{P_1 = P_{atm}} + \underbrace{z_1}_{=h_1} = \frac{V_2^2}{2g} + \underbrace{\frac{P_2}{\rho g}}_{P_2 = P_{atm}} + \underbrace{z_2}_{\substack{=\,0\text{, since the horizontal}\\ \text{references passes}\\ \text{through point }(2)}}$$

The orifice is either at the bottom of the container (see Figure 1.1) or on its side on the surface. The drop in pressure is negligible between points (1) and (2). We obtain $V = \sqrt{2gh}$.

20 Henri Pitot, born in Aramon on May 29, 1695, and who lived until December 27, 1771, was a hydraulic engineer and inventor of the Pitot tube, used to measure the speed of fluids.

21 Evangelista Torricelli, born on October 15, 1608, in Faenza (Italy) and who lived until October 25, 1647, was a physicist and mathematician known in particular for inventing the barometer.

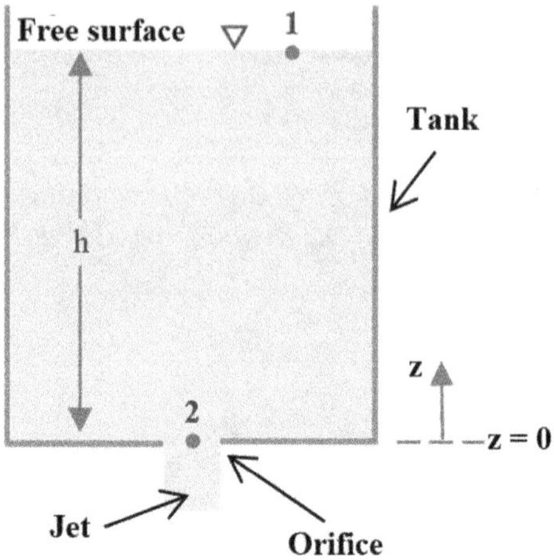

Figure 1.1. *Flow through an orifice, establishing of the Torricelli relationship; HRP: horizontal reference plane. For a color version of this figure, see www.iste.co.uk/sadchemloul/mechanics.zip*

COMMENT 1.3.– *For question 9), relationship* $\tau = \mu \frac{dV}{dr}$ *or Newton's relationship is obtained from the experiment known as "movable wall experiment" given in Figure 1.2.*

Or a fluid located between two horizontal planes separated by a height h. The lower plane is set and the upper plane is movable (in uniform movement) and moves at the speed V_0 *thereby causing the fluid to move.*

Several experiments were carried out using different geometries and fluids of different natures. The results of these experiments show that:

– *the relationship* $\frac{F}{S}$ *is proportional to the relationship* $\frac{V_0}{h}$*;*

– *the proportionality factor depends only on the nature of the fluid;*

– *the relationship between the different physical properties involved in this experiment is written as* $\frac{F}{S} \propto \frac{V_0}{h}$ or $\frac{F}{S} = k\frac{V_0}{h}$*;*

– the relationship $\frac{F}{S}$ *represents the tangential shear stress* $\tau = \frac{F}{S}$. *The triangle formed by* V_0 *and h, and the one formed by dV and dy, are similar right-angled triangles, which makes it possible to write* $\frac{V_0}{h} = \frac{dV}{dy}$ *;*

– the proportionality factor k *depends only on the nature of the fluid and represents dynamic viscosity, which is a thermo-physical property of the fluid; it depends more on temperature than on pressure. The relationship* $\frac{dV}{dy}$ *is called the speed gradient.*

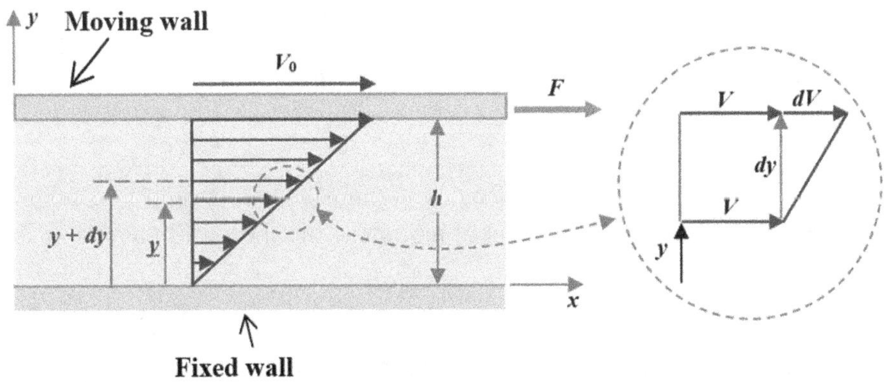

Figure 1.2. *Newton's experiment known as the "moving wall". For a color version of this figure, see www.iste.co.uk/sadchemloul/mechanics.zip*

1.5.2. *Exercise 2: homogeneity of relationships*

1) Check the homogeneity of the relationship $f = \frac{1}{2\pi}\sqrt{\frac{k_r}{m}}$ representing the oscillation frequency f of a solid-spring system, with m the mass of the solid and k_r the spring stiffness constant. The restoring force $\vec{F}$ relates to the extension $\vec{\Delta l}$ with the relationship $\vec{F} = k_r \vec{\Delta l}$.

2) Are the following relationships valid dimensionally? Make a dimensional analysis to confirm or correct this.

a) $F = \frac{Gm}{r}$ like F is a force, G is a constant expressed in $\frac{m^3}{kgs^2}$, m is a mass and r is a distance.

b) $P = \rho g h_1 + h_2 F$ like P is a pressure, g is the acceleration due to gravity, h_1 and h_2 are heights and F force.

c) $\theta = \frac{b\sin(a)}{l\sin(c)}$, l, and c represent lengths.

3) Which of these relationships are homogeneous?

a) $t = 2\pi\sqrt{\frac{l}{g}}$, $t = 2\pi\sqrt{\frac{g}{l}}$, $t = \frac{1}{2\pi}\sqrt{\frac{l}{g}}$, $t = \frac{1}{2\pi}\sqrt{\frac{l+g}{lg}}$

with t as time, l the length and g acceleration from gravity.

b) $E = \sqrt{q_{mv}c^2 + mc^4}$, $E^2 - \frac{q_{mv}^2 c^2}{m} = m^4$, $E^2 = q_{mv}^2 c^2 + m^2 c^4$

with E as energy, q_{mv} the amount of motion, c the speed of light and m the mass.

c) $ma\left(\frac{d\alpha}{dt}\right)^2 = F_N + mg(\sin\alpha) + k_r a(\cos^2\alpha)$

d) $a\left(\frac{d\alpha}{dt}\right)^2 = F_N + g(\sin\alpha) + k_r a(\cos^2\alpha)$

e) $ma\left(\frac{d^2\alpha}{dt^2}\right) = F_N + mg(\sin\alpha) + k_r a(\cos^2\alpha)$

f) $ma\left(\frac{d\alpha}{dt}\right)^2 = F_N + mg(\sin\alpha) + k_r a(\cos\alpha)$

with a being the length, F_N the force, k_r spring stiffness constant, m the mass, t the time and α the angle.

1.5.2.1. *Solutions*

1) Ratio $f = \frac{1}{2\pi}\sqrt{\frac{k_r}{m}}$.

The dimension of the frequency f is $[f] = \left[\frac{1}{\text{time}}\right] = T^{-1}$ and the dimension of the mass is $[m] = M$.

The dimension of the spring stiffness constant k_r is therefore:

$$[k_r] = \frac{[\text{force}]}{[\text{spring extension}]} = \frac{[F]}{[\Delta l]} = MT^{-2}$$

The ratio $\sqrt{\frac{k_r}{m}}$ has the dimension:

$$\left[\frac{k_r}{m}\right]^{1/2} = \frac{[k_r]^{1/2}}{[m]^{1/2}} = \frac{(MT^{-2})^{1/2}}{M^{1/2}} = T^{-1}$$

We recall that the dimension of the constant 2π is equal to the unit.

The relationship $f = \frac{1}{2\pi}\sqrt{\frac{k_r}{m}}$ is therefore homogeneous from the point of view of the dimension.

2) a) The relationship $F = \frac{Gm}{r}$.

The dimension of the force F is $[F] = MLT^{-2}$.

The dimension of G is $[G] = \frac{[\text{lenght}]^3}{[\text{mass}][\text{time}]^2} = M^{-1}L^3T^{-2}$.

The dimension of the relationship $\frac{Gm}{r}$ is:

$$\left[\frac{Gm}{r}\right]=\frac{[G][m]}{[r]}=\frac{M^{-1}L^3T^{-2}M}{L}=L^2T^{-2}$$

which corresponds to the dimension of the square of the speed.

The formula $F=\frac{Gm}{r}$ is not valid dimensionally.

b) relationship $P=\rho gh_1+h_2F$.

The dimension of the pressure is $[P]=ML^{-1}T^{-2}$.

The dimension of the term ρgh_1 is:

$$\left[\rho gh_1\right]=[\rho][g]\left[h_1\right]=ML^{-3}LT^{-2}L=ML^{-1}T^{-2}$$

The dimension of the term h_2F is:

$$\left[h_2F\right]=\left[h_2\right][F]=LMLT^{-2}=ML^2T^{-2}$$

We note here that the terms of the sum ρgh_1+h_2F do not have the same dimension, and therefore the relationship of $P=\rho gh_1+h_2F$ is not valid dimensionally.

c) Ratio $\theta=\frac{b\sin(a)}{l\sin(c)}$.

Since the constants b and l represent lengths, the relationship $\frac{b\sin(a)}{l\sin(c)}$ has no dimension, and since the angle θ has no dimension, then the relationship $\theta=\frac{b\sin(a)}{l\sin(c)}$ is valid dimensionally.

3) a) Ratios $t = 2\pi\sqrt{\frac{l}{g}}$, $t = 2\pi\sqrt{\frac{g}{l}}$, $t = \frac{1}{2\pi}\sqrt{\frac{l}{g}}$, $t = \frac{1}{2\pi}\sqrt{\frac{l+g}{lg}}$.

$$\begin{cases} [t] = T \\ \left[2\pi\sqrt{\frac{l}{g}}\right] = T \end{cases}$$

The relationship $t = 2\pi\sqrt{\frac{l}{g}}$ is homogeneous.

$$\begin{cases} [t] = T \\ \left[2\pi\sqrt{\frac{g}{l}}\right] = T^{-1} \end{cases}$$

The relationship $t = 2\pi\sqrt{\frac{g}{l}}$ is not homogeneous.

$$\begin{cases} [t] = T \\ \left[\frac{1}{2\pi}\sqrt{\frac{l}{g}}\right] = T \end{cases}$$

The relationship $t = \frac{1}{2\pi}\sqrt{\frac{l}{g}}$ is homogeneous.

$$t = \frac{1}{2\pi}\sqrt{\frac{l+g}{lg}}$$

This relationship is not homogeneous, because the terms of a sum must be homogeneous, however l and g do not have the same dimension.

b) $E = \sqrt{q_{mv}c^2 + mc^4}$, $E^2 - \frac{q_{mv}^2 c^2}{m} = m^4$, $E^2 = q_{mv}^2 c^2 + m^2 c^4$.

relationship $E=\sqrt{q_{mv}c^2+mc^4}$:

$$\begin{cases}\left[q_{mv}c^2\right]=\left[q_{mv}\right]\left[c^2\right]=\left[mV\right]\left[c^2\right]=[m][V][c]^2=M(LT^{-1})^3\\ \left[mc^4\right]=\left[m\right]\left[c^4\right]=\left[m\right][c]^4=M(LT^{-1})^4\end{cases}$$

with V being the speed of the mass m.

Since the terms of the sum $q_{mv}c^2+mc^4$ do not have the same dimension, the relationship $E=\sqrt{q_{mv}c^2+mc^4}$ is not homogeneous.

Relationship $E^2-\dfrac{q_{mv}^2c^2}{m}=m^4$:

$$\left[\frac{q_{mv}^2c^2}{m}\right]=\frac{\left[q_{mv}^2c^2\right]}{[m]}=\frac{\left[q_{mv}\right]^2[c]^2}{[m]}=\frac{[mV]^2[c]^2}{[m]}=\frac{(MLT^{-1})^2(LT^{-1})^2}{M}=ML^4T^{-4}$$

$$\left[E^2\right]=M^2LT^{-4}$$

$$\left[m^4\right]=M^4$$

with V being the speed of the mass m. The components of this relationship do not have the same dimension, the relationship $E^2-\dfrac{q_{mv}^2c^2}{m}=m^4$ is not homogeneous.

Relationship $E^2=q_{mv}^2c^2+m^2c^4$:

$$\begin{cases}\left[E^2\right]=M^2L^4T^{-4}\\ \left[q_{mv}^2c^2\right]=M^2L^4T^{-4}\\ \left[m^2c^4\right]=M^2L^4T^{-4}\end{cases}$$

with V being the speed of the mass m.

The relationship $E^2 = q_{mv}^2 c^2 + m^2 c^4$ is homogeneous, because its terms have the same dimension.

c) Relationship $ma\left(\frac{d\alpha}{dt}\right)^2 = F_N + mg(sin\alpha) + ka(cos^2\alpha)$:

$$\begin{cases} \left[ma\left(\frac{d\alpha}{dt}\right)^2\right] = [ma]\left[\left(\frac{d\alpha}{dt}\right)^2\right] = [m][a]\left[\frac{d\alpha}{dt}\right] = MLT^{-2} \\ [F_N] = MLT^{-2} \\ [mg(sin\alpha)] = [m][g] = MLT^{-2} \\ ka(cos^2\alpha) = [k][a] = MT^{-2}L = MLT^{-2} \end{cases}$$

The relationship $ma\left(\frac{d\alpha}{dt}\right)^2 = F_N + mg(sin\alpha) + ka(cos^2\alpha)$ is homogeneous, because all its terms have the same dimension.

d) $a\left(\frac{d\alpha}{dt}\right)^2 = F_N + g(sin\alpha) + ka(cos^2\alpha)$:

$$\begin{cases} \left[a\left(\frac{d\alpha}{dt}\right)^2\right] = [a]\left[\frac{d\alpha}{dt}\right]^2 = LT^{-2} \\ [F_N] = MLT^{-2} \\ [g(sin\alpha)] = LT^{-2} \\ \left[ka(cos^2\alpha)\right] = MLT^{-2} \end{cases}$$

The relationship $a\left(\frac{d\alpha}{dt}\right)^2 = F_N + g(sin\alpha) + ka(\cos^2\alpha)$ is not homogeneous, because its terms do not have the same dimension.

e) $ma\left(\dfrac{d^2\alpha}{dt^2}\right) = F_N + mg(sin\alpha) + ka(cos^2\alpha)$:

$$\begin{cases} \left[ma\dfrac{d^2\alpha}{dt^2}\right] = MLT^{-2} \\ [F_N] = MLT^{-2} \\ [mg(sin\alpha)] = MLT^{-2} \\ \left[ka(cos^2\alpha)\right] = MLT^{-2} \end{cases}$$

The relationship $ma\left(\dfrac{d^2\alpha}{dt^2}\right) = F_N + mg(\sin\alpha) + ka(cos^2\alpha)$ is homogeneous, because its terms have the same dimension.

f) With regard to case 3c), it can easily be verified that the relationship:

$$ma\left(\frac{d\alpha}{dt}\right)^2 = F_N + mg(sin\alpha) + ka(cos\alpha)$$

is homogeneous.

For the four ratios 3c) to 3f), it must be remembered that the dimension of an angle is equal to the unit.

1.5.3. *Exercise 3: dimension of the constants of an equation*

We express the speed of a body by the equation $V = At^3 - Bt$ where t represents time.

1) What are the SI units of A and B ?

2) Give the SI units of A, B and C of the equation $V = At^2 - Bt + \sqrt{C}$, where V is the speed and t is the time.

1.5.3.1. *Solutions*

1) For the equation to be dimensionally valid, the terms V, At^3, and Bt have the same dimension, which must be that of a velocity.

The dimension of A is:

$$[A]=\left[\frac{\text{speed}}{t^3}\right]=\frac{[\text{speed}]}{\left[t^3\right]}=LT^{-4}$$

and its unit is $m.s^{-4}$.

The dimension of B is:

$$[B]=\left[\frac{\text{speed}}{t}\right]=\frac{[\text{speed}]}{[t]}=LT^{-2}$$

and its unit is $m.s^{-2}$, which is that of acceleration.

2) In the same way as 1), we find:

– $[A]=LT^{-3}$ and its unit is $m.s^{-3}$;

– $[B]=LT^{-2}$ and its unit is $m.s^{-2}$;

– $\left[C^{1/2}\right]=[\text{speed}]$, where $[C]=[\text{speed}]^2=L^2T^{-2}$ and its unit is $m.s^{-2}$.

1.5.4. *Exercise 4: equation for perfect gases*

The equation for perfect gases applying to moles is written as $Pv_m=RT$, where P is the pressure, v_m is the molar volume, R is the constant for perfect gases is T and the thermodynamic temperature of the gases.

1) Give the equation for the dimensions of the molar constant of the perfect gases.

2) Knowing that the normal molar volume value is $v_m = 22,414$ *liters / mole*, calculate R using the International System of units.

1.5.4.1. *Solutions*

1) The equation of perfect gases is $Pv_m = RT$, with n being the number of moles, which gives us $v_m = \frac{v}{n}$, which is expressed in *liters/minute* or $m^3/mole$.

With $[R]$ being the size of the constant of perfect gases R and θ the dimension of the temperature T, thus $[P] = ML^{-1}T^{-2}$, $[v_m] = L^3 mole^{-1}$, and $[T] = \theta$.

Therefore, we can write:

$$[R] = [P][v_m][T]^{-1} = ML^2T^{-2}mole^{-1}\theta^{-1}$$

and the unit of the perfect Gas Constant R is therefore $kg.m^2s^{-2}mole^{-1}K^{-1}$.

2) The value of the perfect gas constant is:

$$R = \frac{Pv_m}{T} = \frac{101325 \times 22.414 \times 10^{-3}}{273.5} = 8.314473\ JK^{-1}mole^{-1}$$

1.5.5. *Exercise 5: unit conversions*

1) Convert a force of $1\ N$, an acceleration of $1\ ms^{-2}$ and a dynamic viscosity of $1\ kgm^{-1}s^{-1}$ to CGS units.

2) Convert a pressure of 10 hPa into dynes (CGS units).

1.5.5.1. *Solutions*

1) The dimension of a force F is:

$$[F] = [\text{mass} \times \text{acceleration}] = [\text{mass}] \times [\text{acceleration}] = MLT^{-2}$$

If we consider both systems of units (SI for index 1 and CGS for index 2), we can write:

$$\frac{[F_1]}{[F_2]}=\frac{M_1}{M_2}\frac{L_1}{L_2}\left(\frac{T_1}{T_2}\right)^{-2}$$

The relationship between the units of force taken in two systems is:

$$\frac{\text{unit } SI}{\text{unit } CGS}:\frac{F_1}{F_2}=\frac{1N}{1gcms^{-2}}=\frac{1kg}{1g}\frac{1m}{1cm}\frac{1s^{-2}}{1s^{-2}}=10^3\times10^2\times1=10^5$$

So, we obtain:

$$1\,\text{N}=10^5\,\text{gcms}^{-2}=10\,\text{dyne}$$

The dyne is defined as the force required to accelerate a mass of one gram by $1\,gal$[22] (that is, $1\,cms^{-2}$), or $1\,gcms^{-2}$. A dyne has the value of exactly 10^{-5} newtons.

The dimension of an acceleration value is:

$$[\text{acceleration}]=[\gamma]=LT^{-2}$$

As for acceleration, we can write it as:

$$\frac{[\gamma_1]}{[\gamma_2]}=\frac{L_1}{L_2}\left(\frac{T_1}{T_2}\right)^{-2}$$

The relationship between the units of acceleration taken in two systems is:

$$\frac{\text{unit } SI}{\text{unit } CGS}:\frac{[\gamma_1]}{[\gamma_2]}=\frac{1\,ms^{-2}}{1\,cms^{-2}}=\frac{1m}{1cm}\frac{1s^{-2}}{1s^{-2}}=\times10^2\times1=10^2$$

22 The gal (symbol: Gal) is a CGS unit of acceleration equal to 1 cm/s^2 = 0.01 m/s^2, used to express the acceleration from gravity in geodesy and geophysics.

So, we obtain:

$$1\ ms^{-2} = 10^2\ cms^{-2} = 1\ gal$$

The dimension of dynamic viscosity is:

$$[\mu] = \frac{[\text{tangential stress}]}{[\text{velocity gradient}]} = \frac{[\tau]}{[dV/dy]}$$

$$[\tau] = \frac{[\text{force}]}{[\text{unit area}]} = \frac{MLT^{-2}}{L^2} = ML^{-1}T^{-2}$$

$$\left[\frac{dV}{dy}\right] = \frac{[dV]}{[dy]} = \frac{LT^{-1}}{L} = T^{-1}$$

$$[\mu] = \frac{[\tau]}{[dV/dr]} = ML^{-1}T^{-1}$$

And for force, we can write it as:

$$\frac{[\mu_1]}{[\mu_2]} = \frac{M_1}{M_2}\left(\frac{L_1}{L_2}\right)^{-1}\left(\frac{T_1}{T_2}\right)^{-1}$$

The relationship between the units of viscosity used in two systems is:

$$\frac{\text{unit } SI}{\text{unit } CGS} : \frac{[\mu_1]}{[\mu_2]} = \frac{1\,kgms^{-1}}{1gcms^{-1}} = \frac{1kg}{1g}\frac{1m^{-1}}{1cm^{-1}}\frac{1s^{-1}}{1s^{-1}} = 10^3 \times 10^{-2} \times 1 = 10$$

So, we obtain:

$$1\,kgms^{-1} = 1\text{Poiseuille} = 10\,gcms^{-1} = 10\,Po = 10\text{Poises}$$ [23]

$$1\,PI\ (\text{Poiseuille}) = 1\,Pa.s\ (\text{Pascal.second})$$

23 Jean-Léonard-Marie Poiseuille, born on April 22, 1797, in Paris and who lived until December 26, 1869, was a physicist and doctor who graduated from the École polytechnique (France's Polytechnic University) in 1815. He published several of his findings on the heart and the circulation of blood in vessels.

Poise is the CGS unit of dynamic viscosity, defined as the viscosity of a fluid for which a tangential stress of one dyne/cm^2 allows it to maintain a speed of 1 cm/s between two parallel planes separated by 1 cm of this liquid; a poise has the value of one dyne-second per square centimeter or a tenth of a Poiseuille.

2) The dimension of a measure of pressure is:

$$[P]=\frac{[\text{normal force}]}{[\text{unit area}]}=\frac{MLT^{-2}}{L^2}=ML^{-1}T^{-2}$$

If we consider both systems of units (SI for index 1 and CGS for index 2), we can write:

$$\frac{[P_1]}{[P_2]}=\frac{M_1}{M_2}\left(\frac{L_1}{L_2}\right)^{-1}\left(\frac{T_1}{T_2}\right)^{-2}$$

The relationship between the units of pressure taken in two systems is:

$$\frac{\text{unit } SI}{\text{unit } CGS}:\frac{[P_1]}{[P_2]}=\frac{1\,kgm^{-1}s^{-2}}{1\,gcm^{-1}s^{-2}}=\frac{1kg}{1g}\frac{1\,m^{-1}}{1\,cm^{-1}}\frac{1\,s^{-2}}{1\,s^{-2}}=10^3\times10^{-2}\times1=10$$

So, we obtain:

$$1\,Pa=10\,gcm^{-1}s^{-2}=10\underbrace{\frac{gcms^{-2}}{cm^2}}_{=1\,dyne/cm^2}=10\frac{dyne}{cm^2}$$

$$1\,hPa=10^2(10\,gcm^{-1}s^{-2})=10^2\left(10\frac{dyne}{cm^2}\right)=10^3\frac{dyne}{cm^2}$$

The pascal (Pa) is an SI unit, but non-SI units may be used, such as $\frac{\text{dyne}}{\text{cm}^2}$ *which is called the barye*:

– $1\text{ barye (ba)}=1\text{ dyne}/\text{cm}^2$ *and* $1\text{ dyne}/\text{cm}^2=0.1\text{Pa}$;

– *The millimeter of mercury (mmHg) or centimeter of water (cmH_2O)*;

– *The torr*[24]: $1\ \text{torr} = 1\ \text{mmHg} = 1.98066\times13595 = 133.322\ \text{Pa}$;

– *Normal atmosphere*[25]: 1 atm = 101325 Pa;

– *The technical atmosphere*[26]: 1 atm = 98066 Pa;

– *The bar* : 1 *bar* = 10^5 *Pa*[27].

The PSI is the pound (force per square inch), or pound-force per square inch, abbreviated as "PSI" or "lbf/in^2*". In some cases, this unit is wrongly referred to as "pound per square inch" which is a unit of measurement of stress and pressure used in English speaking countries*:

$$1\,psi = \frac{1\ lbf}{(1\ in)^2} = \frac{4.4482\ N}{(0.0254\ m)^2} = 6894.76\,Pa$$

1 lbf = 1 pound-force, an English unit of force (in the International System of units, strength is measured in newtons) and 1 pound = 453.49237g. The value of the pound-force, with respect to the SI, is defined as the product of 0.45349237×9.80665.

24 The torr (symbol: Torr) or millimeter of mercury is a unit of measure for pressure. This unit is not part of the International System of units, in which the unit of pressure is the pascal. The Torr is an abbreviation of the name of Italian physicist and mathematician Evangelista Torricelli, inventor of the mercury column barometer.

25 The atmosphere, according to its primary definition, is the layer of gas that surrounds a planet. However, the term is used with several other meanings and expressions. The atmosphere contains different gases such as pollution or nitrogen. To measure pressure, for instance, we can measure Earth's atmosphere, which is the gaseous layer surrounding the Earth which is known as "air". Dry air consists of 78.08% nitrogen molecules, 20.95% oxygen molecules, 0.93% argon, 0.04% carbon dioxide and traces of other gases. The atmosphere protects life on Earth by absorbing ultraviolet solar radiation, warming the surface by retaining heat (known as the greenhouse effect) and reducing temperature differences between day and night.

26 The technical atmosphere (symbol: at) is an older unit of pressure, defined as the force exerted by a kilogram on a surface of one square centimeter: 1 normalized atmosphere = 1 kilogram-force per square centimeter (kgf/cm^2).

27 The term "Bar" has its origins in ancient Greek, referring to gravity. The bar has the advantage of being close to the atmosphere (average atmospheric pressure on the surface of the sea). Outside of the SI, this unit derives from the barye (1 bar = 10^6 baryes).

2

Dimensional Analysis: Rayleigh Method and Vaschy-Buckingham Method

2.1. Introduction

Dimensional analysis is a practical method that is used to verify the homogeneity of a physical formula through its dimensional equations and to determine the form of an equation on the basis of the hypotheses of the physical quantities on which they are based.

This method is based on the fact that only quantities of the same size can be compared or added. It provides a basis for modeling using models and for studying the effects of scale.

Dimensional analysis is applied to many problems:

– determination of non-dimensional numbers involved in physical phenomena;

– modeling of phenomena using models;

– determination of scale effects.

There are many different fields where dimensional analysis can be applied, including:

– a loss of load (or drop in pressure) in load flow;

– flows on open surfaces;

– aerodynamics, flow resistance;

– the forming and propagation of waves;

– the resistance of materials.

Dimensional analysis can be a very powerful way to obtain formulas, the literal expressions of given physical quantities, from simple intuitive reasoning.

Dimensional analysis is only an intuitive and approximate method: the accuracy of its reasoning and the results or formulas are only guaranteed to one constant.

2.2. Definition of dimensional analysis

Dimensional analysis involves applying the concept that the conditions established by the fact that the numerical value of a physical quantity depends on the system of units that has been chosen, and applying this to physical problems.

Among the set of physical quantities, it is possible to choose a subset from among them made up of the fundamental quantities, such that all other physical quantities can be expressed according to these fundamental quantities.

For mechanical quantities, the most commonly used subset of fundamental quantities are length *L*, mass *M*, and time *T*. These quantities are the basis of the CGS and MKS systems of units (the "International System", or SI).

In general, a mechanical quantity G will have the following characteristics:

$$[G] = M^{\alpha} L^{\beta} T^{\gamma} \qquad [2.1]$$

The numerical value of G depends on the units chosen for the fundamental quantities. The essential applications of dimensional analysis are:

– the optimization of the experimental approach of a problem by reducing the number of parameters that must be varied to ensure the appropriate coverage of the phenomenon;

– the verification of a relationship or equation;

– the conversion of the numerical value of a physical quantity when changing base units.

2.3. The Rayleigh method[1]

Lord Rayleigh devised a variant that was easier to use. If n is the number of physical quantities G_i involved in a physical phenomenon, then we can write the following:

$$f(G_1, G_2 ......, G_i ...., G_n) = 0 \quad [2.2]$$

We will express one of the physical quantities as a function of $(n-1)$ other physical quantities:

$$G_i = f(G_1, G_2, ..., ...G_{n-1}) \quad [2.3]$$

with:

$$G_i = kG_1^{\alpha} G_2^{\beta} ....G_{n-1}^{z} \quad [2.4]$$

The dimension of the relationship [2.4] is:

$$[G_i] = [G_1]^{\alpha} [G_2]^{\beta} ....[G_{n-1}]^{z} \quad [2.5]$$

where $\alpha, \beta, ..., z$ are coefficients to be determined such that the product of the dimensions of $G_1, G_2, ..., G_{n-1}$ is consistent with the unit of G_i.

2.3.1. *Example of application: the period of the swinging of a pendulum*

Calculation of the period T_p of oscillations of a pendulum of length l and mass m in a gravitational field of g (see Figure 2.1).

1 John William Strutt, better known by his title, Lord Rayleigh, was an English physicist (1842–1919) who studied several branches of physics and mechanics (acoustics, optics, electrodynamics, electromagnetism, fluid viscosity, photography).

The swinging of the pendulum is a physical phenomenon, and the physical quantities involved in this phenomenon are the period T_p, the length l, the mass m and acceleration of gravity g. These four physical quantities verify the following relationship:

$$f(T_p, l, m, g) = 0$$

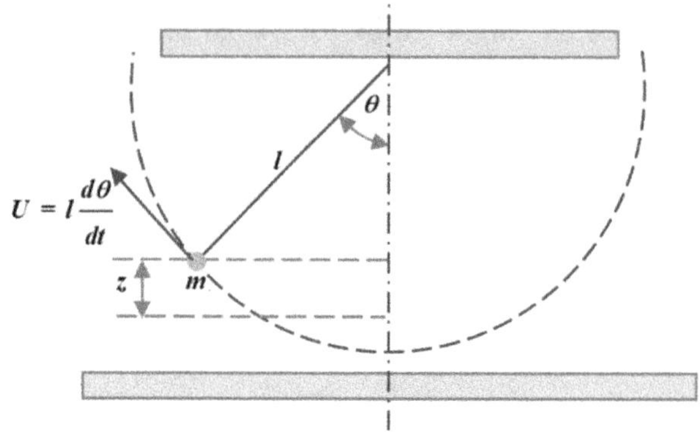

Figure 2.1. *Pendulum in oscillation: z = l(1 – cosθ). For a color version of this figure, see www.iste.co.uk/sadchemloul/mechanics.zip*

By expressing the period T_p according to the other physical quantities, we obtain:

$$T_p = f_1(l, m, g) = kl^{\alpha} m^{\beta} g^{\gamma}$$

The dimension of this relationship gives us:

$$[T] = L^{\alpha} M^{\beta} L^{\gamma} T^{-2\gamma} \text{ or } [T] = M^{\beta} L\alpha^{\alpha+\gamma} T^{-2\gamma}$$

The system of equations obtained is:

$$\begin{cases} M: \beta = 0 \\ L: \alpha + \gamma = 0 \\ T: -2\gamma = 1 \end{cases} \Rightarrow \begin{cases} \alpha = 1/2 \\ \beta = 0 \\ \gamma = -1/2 \end{cases}$$

The equation that gives the period of oscillation of the pendulum is:

$$T_p = k\sqrt{\frac{l}{g}}$$

COMMENT 2.1.– *Solving the equation of the motion of a pendulum gives us a period of* $T_p = k\sqrt{\frac{l}{g}}$, *which is consistent with the result found above.*

– Recall that the equation for the motion of a pendulum is obtained from the conservation of energy, and it is given by:

$$\frac{d^2\theta}{dt^2} = -\frac{g}{l}sin\theta$$

The exact analytical expression of the oscillation period is:

$$\frac{T_p}{T_0} = -\frac{2}{\pi}K\,sin\frac{\theta_0}{2}$$

with $T_0 = 2\pi\sqrt{\frac{l}{g}}$, $\theta_0 = \theta(0)$ *and* K *being a special function called "complete elliptical integral of the first kind". We note that when* $\theta_0 \to 0$, *the period* $T_p \to T_0$ *(see Figure 2.2).*

A pendulum twice as long will have a period $\sqrt{2}$ *times longer, and times* $\sqrt{n}$ *longer for a pendulum* n *times longer exists; this is confirmed through experience.*

Other groupings are possible without dimensions, and this can give other dimensionless products. The most appropriate/suitable ones of these will be found through experience.

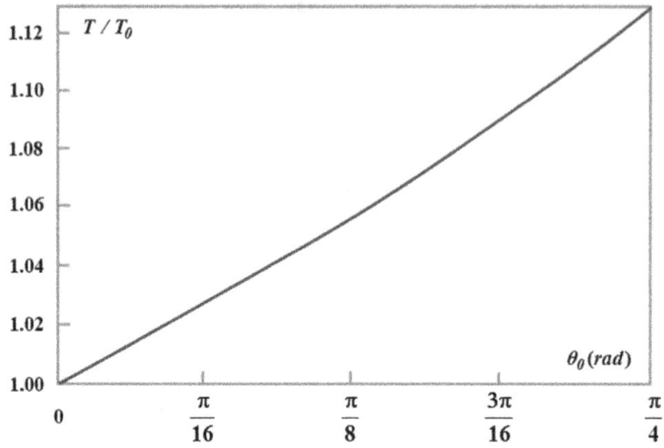

Figure 2.2. *Period of oscillation of a pendulum as a function of the initial angle*

2.4. Vaschy-Buckingham[2] method or method of π

Non-dimensional numbers play an important role. Two questions therefore arise when considering a particular phenomenon:

– how many "dimensional numbers" should we use?

– how can they be determined?

The answer to the first question is given by the Vaschy-Buckingham theorem or the theory of π.

The answer to the second question is more difficult since there are several possible solutions. We need to choose the solutions that best represent the phenomenon: here, we can use certain well-known numbers as guides, such as the Reynolds number[3] and the Froude number[4], which translate

2 Aimé Vaschy, born on April 9, 1857, in Thônes (France) and living until November 25, 1899, was a French telegraph engineer and mathematician. Edgar Buckingham, born on July 8, 1867, in Philadelphia and living until April 29, 1940, was a physicist.

3 Osborne Reynolds, born on August 23, 1842, in Belfast (Northern Ireland) and who lived until February 12, 1912, was an engineer and physicist who made important contributions to hydrodynamics and to fluid dynamics, most notably the creation of the Reynolds number in 1883.

4 William Froude, born November 28, 1810, in Devon (England) and lived until 1879, was an engineer, hydrodynamics specialist, and naval architect.

relationships of power. Then, experience and a good analysis of the phenomenon are indispensable.

2.4.1. *The Vaschy-Buckingham theorem*

We will consider a physical phenomenon whose analysis makes it possible to create a list of physical quantities G_i that have an influence on this phenomenon. We can then assume that the following relationship exists:

$$f(G_1, G_2, ..., G_i, ...G_n) = 0 \qquad [2.6]$$

Each of the physical quantities G_i can be expressed in a single way as a function of $(n-1)$ other quantities.

Or conversely $f(x_1, x_2, ..., x_k)$ a subset of basic quantities sufficient to express the size of all quantities G_i under the form:

$$[G_i] = x_1^{a_{i1}} + x_2^{a_{i2}} + ...x_k^{a_{ik}} \qquad [2.7]$$

It is then shown that products such as:

$$A = G_1^{v_1} G_2^{v_2} ......G_i^{v_i} ......G_n^{v_n} \qquad [2.8]$$

in which v_i are the unknowns and must be dimensionless. Taking into account equation [2.7], the product [2.8] can be written as:

$$A = (x_1^{a_{11}} x_2^{a_{12}} ....x_k^{a_{1k}})^{v_1} (x_2^{a_{21}} x_2^{a_{22}} ....x_k^{a_{2k}})^{v_2} ..... \qquad [2.9]$$

This product is therefore dimensionless if each of the exponents of the fundamental quantities is zero. We thus obtain a system of k algebraic equations containing the n unknowns v_i:

$$\begin{cases} a_{11}v_1 + a_{21}v_2 + ...+ a_{n1}v_n = 0 \\ a_{12}v_1 + a_{22}v_2 + ...+ a_{n2}v_n = 0 \\ .... \\ a_{1k}v_1 + a_{2k}v_k + ...+ a_{nk}v_n = 0 \end{cases} \qquad [2.10]$$

The matrix of the coefficients of the unknowns therefore comprises k rows and columns n :

$$\begin{bmatrix} a_{11} & a_{21} \ldots\ldots\ldots\ldots a_{n1} \\ a_{12} & a_{22} \ldots\ldots\ldots\ldots a_{n2} \\ \ldots\ldots\ldots\ldots\ldots\ldots & \\ a_{1k} & a_{2k} \ldots\ldots\ldots\ldots a_{nk} \end{bmatrix} \qquad [2.11]$$

We will refer to "row r" of this matrix as the number r of rows (or columns) of the largest non-zero determinant that can be deduced from this matrix.

THEOREM 2.1.– *With $f(G_1, G_2, \ldots\ldots, G_i, \ldots G_n)$ being the relationship that exists between n quantities G_i involved in the same phenomenon and k the number of fundamental quantities necessary to express the dimension of G_i.*

If r is the row of matrix [2.10] constructed from the n quantities and of k fundamental quantities, the relationship f can be reduced to a relationship $F(\pi_1, \pi_2, \ldots\ldots, \pi_i, \ldots\ldots, \pi_{n-r})$ between (n – k) products with no dimensions.

COMMENT 2.3.– *It should be noted that very often, row r of the matrix will be equal to the number k of the fundamental quantities, but this is not a general result.*

2.4.2. *Formation of terms in π*

The steps to follow are:

1) identification of n physical quantities of the physical problem;

2) determination of the dimensions of each physical quantity using the basic dimensions (M, L, T);

3) the choice of physical quantities: these quantities must not have the same dimension (which do not form a term in π) and whose number is equal to that of the k fundamental quantities. These selected physical quantities are also called repeated quantities;

4) each term in π is formed by the product of these repeated quantities raised to the powers and of one of the remaining physical quantities raised to the power of one. The number of powers is equal to the number of fundamental quantities;

5) writing the dimension of each term in π, which makes it possible to obtain a system of equations whose unknowns are exponents of repeated physical quantities;

6) determination of exponents to form the terms in π and to find the relationship $F(\pi_1, \pi_2, ...\pi_i, ...\pi_{n-k}) = 0$.

2.4.3. *Application example: linear pressure drop calculation*

We will consider the problem of viscous flow in a round duct for the determination of a dimensionless pressure drop in accordance with other dimensionless parameters. The linear pressure drop ΔP_L depends on the following:

– the flow rate: the average flow rate V;

– the geometric characteristics of the pipe: diameter D, length L_c, the roughness of the inner wall of the pipe ε;

– the properties of the fluid: density ρ of the fluid, dynamic viscosity μ of the fluid.

2.4.3.1. *Solution*

The physical quantities involved in determining the pressure drop verify the equation:

$$f(\Delta P_L, V, D, L_c, \varepsilon, \rho, \mu) = 0$$

By expressing the loss of linear load ΔP_L on the basis of the other physical quantities, we obtain:

$$\Delta P_L = f_1(V, D, L_c, \varepsilon, \rho, \mu)$$

Procedure to follow

Determination of the number of physical quantities $n = 7$.

Determination of the size of the different physical quantities:

$$[\Delta P_L] = ML^{-1}T^{-2},\ [V] = LT^{-1},\ [\varepsilon] = [D] = [L_c] = L$$

$$[\rho] = ML^{-3},\ [\mu] = ML^{-1}T^{-1}$$

The number of fundamental quantities is $k = 3$. Therefore, there are $n - k = 7 - 3 = 4$ terms in π, verifying the following relationship:

$$F(\pi_1, \pi_2, \pi_3, \pi_4) = 0$$

Choice of repeated quantities to form the terms in π: either V, D, and ρ these quantities.

Formation of terms in π:

$$\begin{cases} \pi_1 = V^{\alpha} D^{\beta} \rho^{\gamma} \mu \\ \pi_2 = V^{\alpha} D^{\beta} \rho^{\gamma} L_c \\ \pi_3 = V^{\alpha} D^{\beta} \rho^{\gamma} \varepsilon \\ \pi_4 = V^{\alpha} D^{\beta} \rho^{\gamma} \Delta P_L \end{cases}$$

Determination of the dimension of terms in π:

$$\begin{cases} [\pi_1] = [V]^{\alpha} [D]^{\beta} [\rho]^{\gamma} [\mu] \\ [\pi_2] = [V]^{\alpha} [D]^{\beta} [\rho]^{\gamma} [L_c] \\ [\pi_3] = [V]^{\alpha} [D]^{\beta} [\rho]^{\gamma} [\varepsilon] \\ [\pi_4] = [V]^{\alpha} [D]_{\beta} [\rho]^{\gamma} [\Delta P_L] \end{cases}$$

or:

$$\begin{cases} M^0 L^0 T^0 = (LT^{-1})^{\alpha}(L)^{\beta}(ML^{-3})^{\gamma} ML^{-1}T^{-1} \\ M^0 L^0 T^0 = (LT^{-1})^{\alpha}(L)^{\beta}(ML^{-3})^{\gamma} L \\ M^0 L^0 T^0 = (LT^{-1})^{\alpha}(L)^{\beta}(ML^{-3})^{\gamma} L \\ M^0 L^0 T^0 = (LT^{-1})^{\alpha}(L)^{\beta}(ML^{-3})^{\gamma} ML^{-1}T^{-2} \end{cases}$$

Determination of terms in π.

Term π_1:

$$\begin{cases} M: \gamma + 1 = 0 \\ L: \alpha + \beta - 3\gamma + 1 = 0 \\ T: -\alpha - 1 = 0 \end{cases} \Rightarrow \begin{cases} \alpha = -1 \\ \beta = -1 \\ \gamma = -1 \end{cases} \Rightarrow \pi_1 = \frac{\mu}{\rho V D}$$

This is the inverse of the Reynolds number, but it can be written without loss of generality as:

$$\pi_1 = \frac{\rho V D}{\mu} = Re$$

Term π_2:

$$\begin{cases} M: \gamma = 0 \\ L: \alpha + \beta - 3\gamma + 1 = 0 \\ T: -\alpha = 0 \end{cases} \Rightarrow \begin{cases} \alpha = 0 \\ \beta = -1 \\ \gamma = 0 \end{cases} \Rightarrow \pi_2 \frac{L_c}{D}$$

This is the ratio between the length and diameter of the pipe.

Term π_3:

$$\begin{cases} M: \gamma = 0 \\ L: \alpha + \beta - 3\gamma + 1 = 0 \\ T: -\alpha = 0 \end{cases} \Rightarrow \begin{cases} \alpha = 0 \\ \beta = -1 \\ \gamma = 0 \end{cases} \Rightarrow \pi_3 = \frac{\varepsilon}{D}$$

This is the relative roughness, which is the ratio of the absolute roughness of the pipe to its diameter.

Term π_4 :

$$\begin{cases} M:\gamma+1=0 \\ L:\alpha+\beta-3\gamma+1=0 \\ T:-\alpha-2=0 \end{cases} \Rightarrow \begin{cases} \alpha=2 \\ \beta=0 \\ \gamma=-1 \end{cases} \Rightarrow \pi_4=\frac{\Delta P_L}{\rho V^2}$$

This term is equal to half of the pressure coefficient, which is the ratio of the drop in pressure to the dynamic pressure. Dimensional analysis makes it possible to write the following:

$$\pi_4=\frac{\Delta P_L}{\rho V^2/2}$$

The terms π verify the following relationship:

$$F(\pi_1,\pi_2,\pi_3,\pi_4)=0 \text{ or } F(\frac{\rho VD}{\mu},\frac{L_c}{D},\frac{\varepsilon}{D},\frac{\Delta P_L}{\rho V^2/2})=0$$

The expression of the loss of linear pressure is given by:

$$\frac{\Delta P_L}{\rho V^2/2}=F_1(\frac{\rho VD}{\mu},\frac{L_c}{D},\frac{\varepsilon}{D})$$

COMMENT 2.4.– *The linear head drop or the linear head loss is therefore a function of the Reynolds number, the ratio between the length and diameter of the pipe, and the relative roughness of the pipe (ratio between the absolute roughness and the diameter of the pipe). The absolute roughness is an indication of the state of the inner surface of the pipe. It depends on the nature of the material used to make the pipe.*

The relationship of the linear pressure drop would be different if the repeated quantities were different than those considered in this application. Dimensional analysis is an extremely powerful tool, but it is also very dangerous. If we forget about or make a mistake in the choice of physical variables to be considered, the results are then false. The physical sense must allow the selection of the relevant independent variables.

2.5. Exercises: homogeneity method or Rayleigh method

2.5.1. *Exercise 1: Reynolds number*

The Reynolds number Re is a function of the density ρ of the fluid, the flow velocity V of the fluid. and a characteristic length L_c.

2.5.1.1. *Solution*

The relationship between the Reynolds number and other physical quantities is:

$$Re = f(V, \rho, \mu, L_c) = kV^{\alpha} \rho^{\beta} \mu^{\gamma} L_c^{a}$$

with k being the proportionality constant.

The dimension of this expression gives us:

$$M^0 L^0 T^0 - L^{\alpha} T^{-\alpha} M^{\beta} L^{-3\beta} M^{\gamma} L^{\gamma} T^{-\gamma} L^{a} \Rightarrow \begin{cases} M : \beta + \gamma \\ L : \alpha - 3\beta - \gamma + a = 0 \\ T : - \alpha - \gamma = 0 \end{cases}$$

This system of equations is incompatible, since the number of equations is less than the number of unknowns. Thus, we must express the three exponents, α , β , and γ as a function of a , which gives us:

$$\begin{cases} \alpha = - a \\ \beta = - a \\ \gamma = - a \end{cases}$$

The expression of the Reynolds number is written as:

$$Re = k\left(\frac{\mu}{\rho V L_c}\right)^{a}$$

The values of k and a are determined experimentally.

COMMENT 2.5.– *The values of* k *and* a *are determined experimentally. The exponent* γ *depends on rheological behavior*[5]*: Newtonian laws, plastics, pseudo-plastics, etc. For* $k=1$ *and* $a=-1$, *we may write in a general sense*:

$$Re = \frac{\rho V_R L_R}{\mu}$$

V_R *and* L_R *are the reference speed and the reference length, respectively.*

The Reynolds number represents the ratio of the forces of inertia to viscous forces. In the case of a pipe with a circular cross-section, the characteristic length, also known as the reference length, represents the diameter of this section.

If the straight section of the pipe is not circular, the hydraulic diameter shall be used in calculating the Reynolds number. The hydraulic diameter is given by:

$$D_H = \frac{\text{4 times the cross-sectional area of the flow}}{\text{wetted perimeter of the cross-section}} = \frac{4S_m}{Pe_m}$$

Table 2.1 shows the expression of the hydraulic diameter of sections with simple geometry for load flows and free surface flows. The reference length can thus represent a height, width, length, thickness, etc.

The Reynolds number allows us to determine the nature of the flow type: the laminar type $(\text{Re} < 2100)$, *the transition type* $(\text{Re} = 2100)$, *and the turbulent type* $(\text{Re} > 2100)$.

Rayleigh's method provides the form of the law that determines the physical phenomenon (however, there is no evidence that the corresponding physical law exists; this is confirmed through experience). Other groupings are possible without dimensions, and this can give other dimensionless products. The most appropriate/suitable ones of these will be found through experience.

5 Rheology (which derives from the Greek word *rheo*, meaning "flow", and *logos*, referring to "study") is the study of the deformation and flow of matter under the effect of applied stress. The word *rheology* was first used in 1928 by Eugene Cook Bingham (born on December 8, 1878, in Vermont (United States), and who lived until November 6, 1945); he was a professor at Lehigh University in the United States.

	Free surface flow	Load flows
Shape of cross section	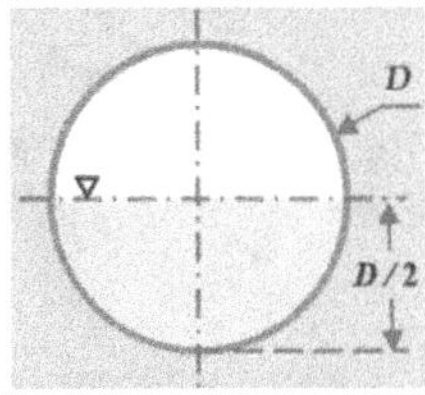	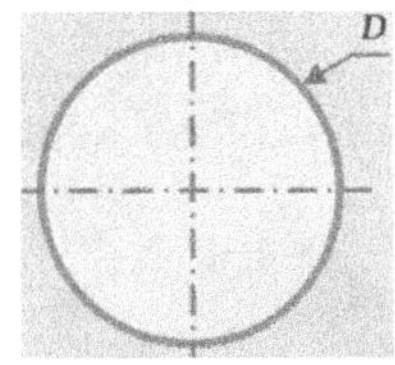
Cross-section area	$S_m = \frac{1}{2}\frac{\pi D^2}{4}$	$S_m = \frac{\pi D^2}{4}$
Wetted perimeter	$Pe_m = \frac{\pi D}{2}$	$Pe_m = \pi D$
Hydraulic diameter	$D_H = \frac{4S_m}{Pe_m} = D$	$D_H = \frac{4S_m}{Pe_m} = D$
Shape of cross section	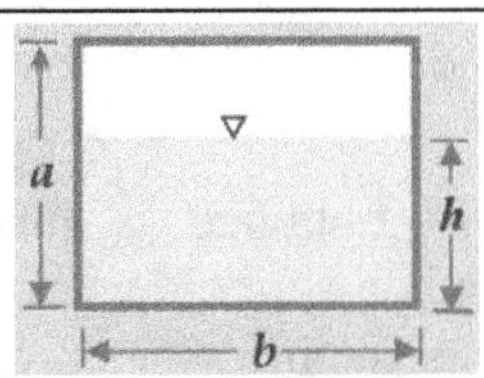	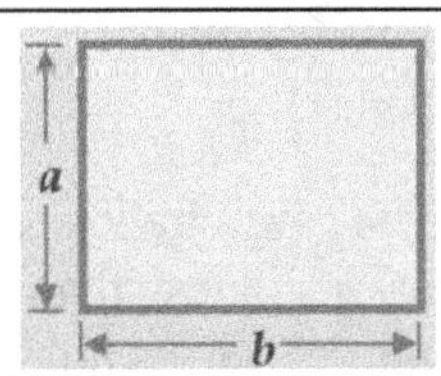
Cross-section area	$S_m = bh$	$S_m = ab$
Wetted perimeter	$Pe_m = b + 2h$	$Pe_m = 2(b + a)$
Hydraulic diameter	$D_H = \frac{4bh}{b+2h}$	$D_H = \frac{4bh}{2(a+b)}$

Table 2.1. *Some expressions of hydraulic cross-section diameter using simple geometry. For a color version of this figure, see www.iste.co.uk/sadchemloul/mechanics.zip*

2.5.2. *Exercise 2: the Weber number*

Establish the expression for the Weber number We, assuming it is a function of the velocity V, the density ρ, the length L_c, and the surface tension σ.

2.5.2.1. *Solution*

The expression of the Weber number We is written as:

$$We = f(\rho, V, L_c, \sigma) = k\rho^{\alpha} V^{\beta} L_c^{\gamma} \sigma^{a}$$

with k being the proportionality constant.

Given that this number is dimensionless, the dimension of this expression gives:

$$M^0 L^0 T^0 = (ML^{-3})^{\alpha} (LT^{-1})^{\beta} L^{\gamma} (MT^{-2})^{a} \Rightarrow \begin{cases} \alpha + a = 0 \\ -3\alpha + \beta + \gamma + 1 = 0 \\ -\beta - 2a = 0 \end{cases}$$

The system of equations obtained is incompatible, the number of equations is less than the number of unknowns, expressing the exponents α, β, and γ, depending on a, we obtain:

$$\begin{cases} \alpha = -\,a \\ \beta = -2a \\ \gamma = -\,a \end{cases}$$

The expression of the Weber number is written as:

$$We = k\left(\frac{\sigma}{\rho V^2 L_c}\right)^{a}$$

The constant k and the exponent a determined experimentally. For $k = 1$ and $a = -1$, the Weber number is written as:

$$We = \frac{\rho V^2 L_c}{\sigma}$$

COMMENT 2.6.– *The Moritz Weber*[6] *number is a dimensionless number used in fluid mechanics to characterize the flow of fluids at the interface points of a multiphase system. It represents the ratio of inertial forces and surface tension, and thus it is written as:*

$$We = \frac{\rho V^2 L_c}{\sigma}$$

It represents the relative importance of shearing forces and surface tension.

Weber's number is primarily used for the study of the flow of films and the formation of drops or bubbles. For example, if a drop has a Weber value of greater than 12, it will disintegrate into many other small drops.

An example of the application of the Weber number is an impact on a liquid film: when an aerosol hits a surface, a liquid film may be created. Therefore, after a certain period of time, the aerosol no longer directly hits the solid surface, but a liquid film. This situation then changes the conditions of impact. This scenario is studied particularly in the context of internal combustion engines in order to model the interactions between the fuel that is injected and the cylinder walls.

Table 2.2 shows various impact regimes.

	Nature of the regime	
Expression of the Weber number	**The drop is captured**	**Partial coalescence regime**
	$We < 5$	$5 < We < 10$
Expression of the Weber number	**Coalescence regime**	**Splash regime**
	$10 < We < 18\frac{\rho v_l f^{3/4}}{\sigma} d_p^2$	$We > 18\frac{\rho v_l f^{3/4}}{\sigma} d_p^2$

Table 2.2. *Various impact regimes of drops on a liquid film*

6 Moritz Weber, born on July 18, 1871, in Leipzig (Germany), and who lived until June 10, 1951, was a fluid dynamics engineer.

In this table, ρ and ν_l are the density and kinetic viscosity of the liquid, d_p is the diameter of the drop, and f is the frequency of the impacting drops, respectively.

2.5.3. *Exercise 3: capillary number*

Establish the expression of the dimensionless capillary number C_a as a function of gravitational acceleration μ, surface tension, σ and velocity d_p.

2.5.3.1. *Solution*

The expression of the capillary number is written as:

$$C_a = f(\mu, \sigma, V) = k\mu^{\alpha}\sigma^{\beta}V^{\gamma}$$

with k being the proportionality constant.

The dimension of this expression lets us write:

$$M^0L^0T^0 = (ML^{-1}T^{-1})^{\alpha}(MT^{-2})^{\beta}(LT^{-1})^{\gamma} \Rightarrow \begin{cases} M: \alpha + \beta = 0 \\ L: -\alpha + \gamma = 0 \\ T: -\alpha - 2\beta - \gamma = 0 \end{cases}$$

For the solution of this system, the exponents α and β are expressed as functions of γ. The third equation of the system makes it possible to verify that the solutions found are exact.

The solutions to the system are:

$$\begin{cases} \alpha = \gamma \\ \beta = -\gamma \end{cases}$$

So, we obtain:

$$C_a = k\left(\frac{\mu V}{\sigma}\right)^{\gamma}$$

The constant k and the exponent α are determined experimentally. For $k=1$ and $\gamma=1$, the capillary number is written as $C_a = \frac{\mu V}{\sigma}$.

REVIEW.– The capillary number is a dimensionless number used in fluid mechanics. It represents the ratio of viscous forces to surface tension and is used to characterize the atomization of liquids.

This number is used to evaluate the effects of surface tension, such as the study of using the propulsion of fluid film by pulling a plate or the study of dynamic angles of dampening.

The capillary number is also utilized in the rheology of fluids containing bubbles or drops (such as emulsions), to determine if these inclusions become distorted during flows.

If $C_a \ll 1$, the tension effects overpower the viscous forces, and if $C_a \gg 1$, the viscous forces overpower the tension effects.

The viscosity is so great that the effects of surface tension on the open surface are negligible.

2.5.4. *Exercise 4: power of a propeller*

Assuming that the power $p_{propeller}$ exerted by a propeller is a function of the density ρ and dynamic viscosity μ of the air, of the diameter D and rotation speed N of the propeller, the speed V of the air current, and the speed of sound c, find this expression.

2.5.4.1. *Solution*

The power $p_{propeller}$ exerted by the propeller therefore depends on:

– the geometric characteristics of the propeller: diameter D and speed N;

– the properties of the fluid: density ρ and dynamic viscosity μ;

– the speed V of the air;

– the speed of sound c.

The expression of the power exerted by the propeller is:

$$p_{propeller} = kD^{\alpha} N^{\beta} \rho^{\gamma} \mu^{\theta} V^{a} c^{b}$$

with k being the proportionality constant.

The dimension of this expression gives us:

$$ML^2T^{-3} = L^{\alpha}T^{-\beta}(ML^{-3})^{\gamma}(ML^{-1}T^{-1})^{\theta}(LT^{-1})^{a}(LT^{-1})^{b}$$

that is:

$$\begin{cases} M: \gamma + \theta = 1 \\ L: \alpha - 3\gamma - \theta - a - b = 2 \\ T: \beta + \theta + a + b = 3 \end{cases}$$

The system of equations obtained is incompatible, so we will express the exponents α, β, and γ as a function of θ, a and b, that is:

$$\begin{cases} \alpha = 5 - 2\theta - a - b \\ \beta = 3 - \theta - a - b \\ \gamma = 1 - \theta \end{cases}$$

The expression of the power generated by the propeller is:

$$p_{propeller} = kD^{5-2\theta-a-b} N^{3-\theta-a-b} \rho^{1-\gamma} \mu^{\theta} V^{a} c^{b}$$

$$= k\rho D^5 N^3 \left(\frac{\rho N D^2}{\mu}\right)^{-\theta} \left(\frac{DN}{V}\right)^{-a} \left(\frac{DN}{c}\right)^{-b}$$

The constant k and exponents θ, a and b are determined experimentally.

For $k = 1$ and $\theta = a = b = -1$, we obtain:

$$p_{propeller} = \rho N^3 D^5 \left(\frac{\rho N D^2}{\mu}\right)\left(\frac{DN}{V}\right)\left(\frac{DN}{c}\right)$$

COMMENT 2.7.– *The term* $\frac{\rho N D^2}{\mu}$ *is proportional to the Reynolds number, and it can effectively be written as:*

$$\frac{\rho N D^2}{\mu} = \frac{60}{\pi} \frac{\rho \omega (D/2) D}{\mu} = \frac{60}{\pi} \frac{\rho V D}{\mu} = \frac{60 Re}{\pi}$$

The term $\frac{DN}{V}$ *defines the pitch of the propeller, that is, the forward movement of the propeller after each turn. It can be written as:*

$$\frac{DN}{V} = \frac{60}{\pi} \frac{\omega D}{2} \frac{1}{V} = \frac{60}{\pi} \frac{\omega D / 2}{V} = \frac{60}{\pi} \frac{\text{linear speed of propeller}}{\text{propeller advance speed}}$$

That is:

$$\frac{DN}{V} = \frac{60}{\pi} \frac{\omega D}{2} \frac{1}{V} = \frac{60}{\pi} \underbrace{\frac{\omega D / 2}{V}}_{\text{propeller pitch}} = \frac{60}{\pi} P_{\text{propeller pitch}}$$

The term $\frac{DN}{c}$ *is proportional to the Mach number*[7]*, and it can effectively be written as:*

$$\frac{DN}{c} = \frac{60}{\pi} \frac{\omega D / 2}{c} = \frac{60}{\pi} \frac{V}{c} = \frac{60}{\pi} Ma$$

The power exerted by a propeller is therefore:

$$P_{\text{propeller}} = \rho N^3 D^5 f(Re, Ma, P_{\text{propeller pitch}})$$

Figure 2.3 below shows an example of a propeller.

7 Ernst Mach, born on February 18, 1838, in Chiltlitz-turas (Austria) and who lived until February 19, 1916, was a physicist and philosopher.

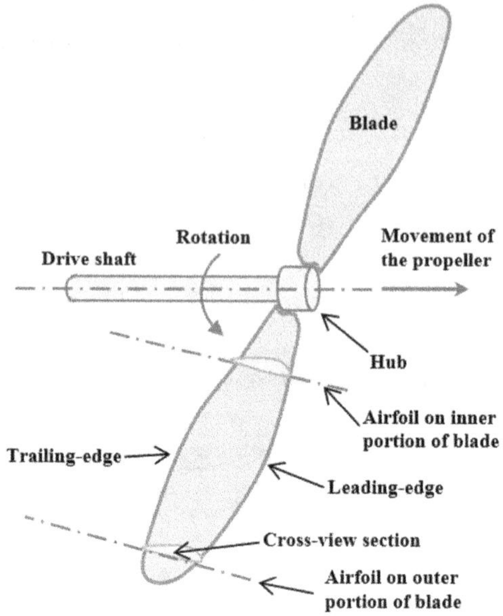

Figure 2.3. *Diagram of an example of a propeller*

NOTE 2.1.– *A propeller is made up of at least two blades joined together around a hub, which itself is connected to a drive shaft. It can easily be noticed (see Figure 2.3) that each blade, shown in a cross section, has very obvious similarities with the wing of an airplane.*

The energy output generated by the motor is transformed into forward propulsion along a straight line. Thanks to its aerodynamic properties, the propeller transforms the torque provided by the motor into a force that provides for the forward movement of the airplane through the air.

2.5.5. *Exercise 5: flow through an orifice with thin walls*

For a perfect liquid, express the volumetric flow Q_v through a thin-walled orifice, depending on the density ρ of the liquid, the diameter d_o of the orifice and the difference in pressure ΔP.

2.5.1.1. *Solution*

Consider a closed tank with an orifice as shown in Figure 2.4:

$$Q_v = f(\rho, d_o, \Delta P) = k\rho^{\alpha} d_o^{\beta} \Delta P^{\gamma}$$

where k is the proportionality constant.

The dimension of this relationship of flows is:

$$L^3T^{-1} = (ML^{-3})^{\alpha}(L)^{\beta}(ML^{-1}T^{-2})^{\gamma} \Rightarrow \begin{cases} M: \alpha + \gamma = 1 \\ L: -3\alpha + \beta - \gamma = 3 \\ T: 2\gamma = 1 \end{cases}$$

The solutions of the systems of equations representing the values of the exponents are:

$$\begin{cases} \alpha = -1/2 \\ \beta = 2 \\ \gamma = 1/2 \end{cases}$$

The flow through a thin-walled orifice is:

$$Q_v = kd_o^2\sqrt{\frac{\Delta P}{\rho}}$$

COMMENT 2.8.– *On the basis of the shape of the edges of the orifice cross section* S_o *(see Figure 2.4), we may draw a distinction between thin-walled orifices with sharp edges (see Figure 2.4a) and round-edged orifices or orifices which are machine-shaped so that the flow is properly guided by the walls (see Figure 2.4b).*

The axis of the jet continues in its initial direction, but due to viscous or turbulent friction, the jet spreads out, and the velocity downstream decreases. On the basis of the contracted section or the vein S_v *of the liquid, the trajectories become parallel to each other, and the pressure distribution becomes hydrostatic in nature.*

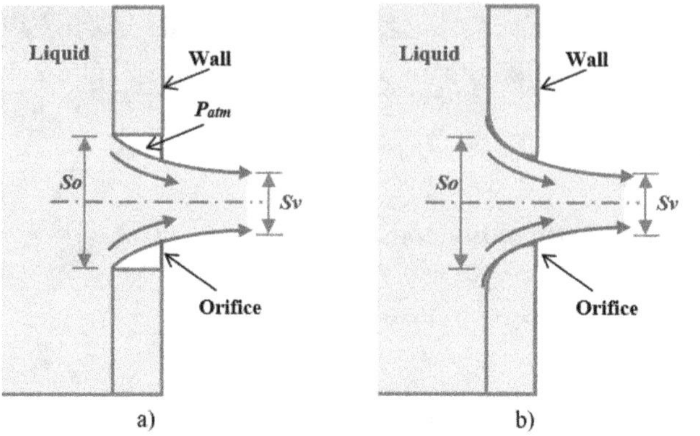

Figure 2.4. *Flow through orifices: (a) a thin-walled orifice; (b) a round-edged orifice. For a color version of this figure, see www.iste.co.uk/sadchemloul/mechanics.zip*

2.5.6. *Exercise 6: a linear pressure drop along a horizontal pipe*

Establish the expression of ΔP_L, a linear pressure drop for an isovolume turbulent flow within a horizontal pipe. This pressure drop depends on the diameter D and length L_c of the pipe, the roughness ε of the inner surface of the pipe, the density ρ and dynamic viscosity μ of the fluid, and the velocity V of the flow of the fluid in the pipe.

2.5.6.1. *Solution*

The relationship between the drop in pressure and other physical quantities is:

$$\Delta P_L = f(\underbrace{D, L_c, \varepsilon}_{\text{characteristics of the pipe}} ; \underbrace{\rho, \mu}_{\text{properties of the fluid}} ; \underbrace{V}_{\text{velocity of the fluid flow}}) = kD^{\alpha} L_c^{\beta} \varepsilon^{\gamma} \rho^{\theta} \mu^{a} V^{b}$$

$$ML^{-1}T^{-2} = (L)^{\alpha}(L)^{\beta}(L)^{\gamma}(ML^{-3})^{\theta}(ML^{-1}T^{-1})^{a}(LT^{-1})^{b}$$

That is:

$$\begin{cases} M: \theta + a = 1 \\ L: \alpha + \beta + \gamma - 3\theta - a + b = -1 \\ T: a + b = 2 \end{cases}$$

This system of equations is incompatible (three equations with six unknowns); the unknowns (exponents) θ, a, and b are determined based on the unknowns (exponents) α, β, and γ which are considered as fixed:

$$\begin{cases} \theta = 1 + (\alpha + \beta + \gamma) \\ a = -\ (\alpha + \beta + \gamma) \\ b = 2 + (\alpha + \beta + \gamma) \end{cases}$$

The relationship of the linear pressure drop is written as:

$$\Delta P_L = kD^{\alpha} L_c^{\beta} \varepsilon^{\gamma} \rho^{1+(\alpha+\beta+\gamma)} \mu^{-(\alpha+\beta+\gamma)} V^{2+(\alpha+\beta+\gamma)}$$

$$= k\rho V^2 \left(\frac{\rho VD}{\mu}\right)^{\alpha} \left(\frac{\rho VL_c}{\mu}\right)^{\beta} \left(\frac{\varepsilon\rho V}{\mu}\right)^{\gamma}$$

COMMENTS 2.9.–

The term $\left(\frac{\rho VD}{\mu}\right)^{\alpha}$ *represents the Reynolds number*:

$$\left(\frac{\rho VD}{\mu}\right)^{\alpha} = Re^{\alpha}$$

The term $\left(\frac{\rho L_c V}{\mu}\right)^{\beta}$ *is written as:*

$$\left(\frac{\rho L_c V}{\mu}\right)^{\beta} = \left(\frac{\rho DV}{\mu}\right)^{\beta} \left(\frac{L_c}{D}\right)^{\beta} = Re^{\beta} \left(\frac{L_c}{D}\right)^{\beta}$$

The term $\left(\frac{\varepsilon\rho V}{\mu}\right)^{\gamma}$ *is written as:*

$$\left(\frac{\varepsilon\rho V}{\mu}\right)^{\gamma} = \left(\frac{\rho VD}{\mu}\right)^{\gamma} \left(\frac{\varepsilon}{D}\right)^{\gamma} = Re^{\gamma} \left(\frac{\varepsilon}{D}\right)^{\gamma}$$

The expression of the loss of pressure is thus given by:

$$\Delta P_L = kRe^{(\alpha+\beta+\gamma)}\left(\frac{\varepsilon}{D}\right)^{\gamma}\rho V^2\left(\frac{L_c}{D}\right)^{\beta}$$

The term $Re^{(\alpha+\beta+\gamma)}\left(\frac{\varepsilon}{D}\right)^{\gamma}$ defines the coefficient of linear pressure drop λ:

$$\lambda = Re^{(\alpha+\beta+\gamma)}\left(\frac{\varepsilon}{D}\right)^{\gamma} = f(Re,\frac{\varepsilon}{D})$$

which gives:

$$\Delta P_L = kf(Re,\frac{\varepsilon}{D})\rho V^2\left(\frac{L_c}{D}\right)^{\beta} = k\lambda\rho V^2\left(\frac{L_c}{D}\right)^{\beta}$$

By taking $\alpha = \beta = \gamma = 1$ and $k = \frac{1}{2}$, we obtain the Darcy-Weisbach relationship[8], which gives the linear pressure drop:

$$\Delta P_L = \lambda\frac{\rho V^2 L_c}{2D} = f(Re,\frac{\varepsilon}{D})\frac{\rho V^2 L_c}{2D}$$

The coefficient λ is determined either using the ratios given in Table 2.3 or using the Moody chart[9] (see Figure 2.5).

Table 2.3 gives the expressions of the coefficient of the linear pressure drop as a function of the Reynolds number and relative roughness.

8 Henry Philibert Gaspard Darcy, born on June 10, 1803, in Dijon (France), and who lived until January 2, 1858, was a hydraulicist. Julius Ludwig Weisbach, born on August 10, 1806, in Mittelschmiedeberg (Germany), and who lived until February 24, 1871, was a mathematician and engineer.

9 Lewis Ferry Moody, born on January 5, 1880, in Philadelphia and who lived until February 21, 1953, was an engineer and professor, best known for devising the Moody chart.

Regime	*Re* value	Expression of λ	Relationship name
Laminar	Re < 2000	$\lambda = \frac{64}{Re}$	Poiseuille[10]
Smooth turbulent	$2000 < \text{Re} < 10^5$	$\lambda = \frac{0.316}{Re^{0.25}}$	Blasius[11]
Smooth turbulent	$10^5 < \text{Re} < 10^6$	$\frac{1}{\sqrt{\lambda}} = 2\log\left(Re\sqrt{\lambda}\right) - 0.8$	Prandtl-Nikuradse[12]
Rough Turbulent	$\text{Re} > 10^6$	$\frac{1}{\sqrt{\lambda}} = -2\log\left(\frac{\varepsilon}{3.71D}\right)$	Von Kármán[13]-Nikuradse
Turbulent	–	$\frac{1}{\sqrt{\lambda}} = -2\log\left(\frac{\varepsilon}{3.71D} + \frac{2.51}{Re\sqrt{\lambda}}\right)$	Colebrook[14]

Table 2.3. *Expression of the coefficient of linear pressure drop* $\lambda = f(\text{Re}, \varepsilon/D)$

COMMENT 2.10.– *The Moody chart is the representation on a single diagram of the general expression of the coefficient of the drop in pressure*:

$$\lambda = f(Re, \frac{\varepsilon}{D})$$

10 Jean-Léonard-Marie Poiseuille, born on April 22, 1797, in Paris and who lived until December 26, 1869, was a physicist and doctor who graduated from the École polytechnique (France's Polytechnic University).

11 Paul Richard Heinrich Blasius, born on August 9, 1883 in Paris, and who lived until April 24, 1970, was a French physicist who worked on fluid dynamics. He was one of the first students of Ludwig Prandtl.

12 Ludwig Prandtl, born on February 4, 1875, in Freising, Germany, and who lived until August 15, 1953, was an engineer and physicist. Ivan Nikuradse, born on November 20, 1894, in Samtredia, in the country of Georgia, and who died on July 18, 1979, was an engineer and physicist.

13 Theodore Von Kármán, born on May 11, 1881, in Budapest (Hungary), and who lived until May 6, 1963, was an engineer and physicist who specialized in aeronautics.

14 Cyril Frank Colebrook, born on July 26, 1910, in Swansea, Wales, and who lived until January 12, 1997, was a physicist. This relationship very correctly reflects the behaviors of industrial pipes, for which the functionality schemes often correspond to the domain of transition.

The chart was established on the basis of the theoretical studies done by Blasius and Prandtl, the numerous and well-known experiments carried out by Nikuradse (on variable roughness) and measurements made on industrial pipes.

The universal Moody diagram is shown in Figure 2.5. It provides the curves $\lambda = f(Re)$ for each value of the relative roughness $(\frac{\varepsilon}{D})$.

Curve A represents the relationship of Poiseuille under a laminar regime (see Table 2.3).

Curves B and C represent the relationship of Blasius and Prandtl-Nikuradse when the turbulence regime is smooth (see Table 2.3).

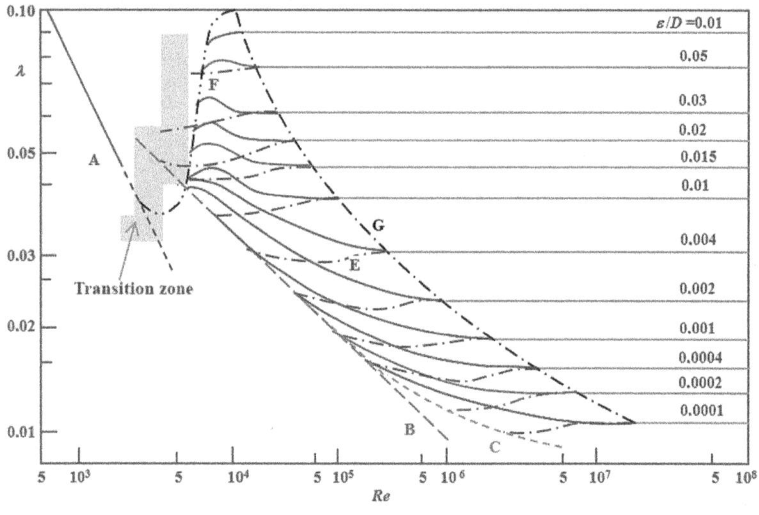

Figure 2.5. *Moody diagram* $\lambda = f(Re, \frac{\varepsilon}{D})$

The area between curves C and G is the transition zone between hydraulically smooth and hydraulically rough flows. In this area, the coefficient of the linear pressure drop λ depends on both the Reynolds number Re *and the relative roughness $(\frac{\varepsilon}{D})$.*

The E curves, known as Nikuradse's harp, were obtained experimentally by Nikuradse by varying both the diameter and the roughness of the pipe.

The roughness was obtained by gluing grains of sand with a well-defined granular size onto the walls of the pipe. This allowed for a uniform and very regular roughness level, both in size and distribution.

This explains the difference found in the F curves obtained by Colebrook from experiments with industrial pipes, for which roughness can vary widely. He made important contributions to fluid mechanics. He is best known for the table named after him, which provides the roughness of pipes.

Table 2.4 shows some common values of absolute roughness.

Nature of the inner surface of the pipe	Absolute roughness ε (mm)
PVC	0.0015
Stainless steel	0.015
Commercial steel	0.045 to 0.09
Drawn steel	0.015
Welded steel	0.045
Galvanized steel	0.15
Rusted steel	0.1 to 1
New pig iron	0.25 to 0.8
Used cast iron	0.8 to 1.5
Cast iron	1.5 to 2.5
Sheet or asphalt cast iron	0.01 to 0.015
Well-smoothed cement	0.3
Reinforced concrete	1
Rough concrete	5

Table 2.4. *Standard values of absolute roughness (ε) in mm*

2.5.7. Exercise 7: force exerted by a fluid on a body

Given that the resistance F exerted by a moving fluid on a body is a function of the density ρ of the fluid, the dynamic viscosity μ of the fluid, the velocity at infinity V_∞, the pressure at infinity P_∞ of the fluid, and a characteristic length L_c of the body, establish the general equation of F.

2.5.7.1. *Solution*

The relationship between the strength of the resistance F and other physical quantities is written:

$$F = f(\rho, \mu, V_\infty, P_\infty, L_c) = k\rho^\alpha \mu^\beta V_\infty^\gamma P_\infty^\theta L_c^a$$

The dimension of this relationship is:

$$MLT^{-2} = (ML^{-3})^{\alpha}(ML^{-1}T^{-1})^{\beta}(LT^{-1})^{\gamma}(ML^{-1}T^{-2})^{\theta}(L)^{a}$$

$$\begin{cases} \alpha+\beta+\theta=1 \\ -3\alpha-\beta+\gamma-\theta+a=1 \\ -\beta-\gamma-2\theta=-2 \end{cases} \Rightarrow \begin{cases} \alpha=1-\theta-a \\ \beta=2-\theta \\ \gamma=2-\theta-2a \end{cases}$$

The expression of the resistance force is therefore:

$$F = k\rho V_{\infty}^{2} L_{c}^{2}\left(\frac{\mu}{\rho V_{\infty} L_{c}}\right)^{\theta}\left(\frac{P_{\infty}}{\rho V_{\infty}^{2}}\right)^{a}$$

NOTE 2.2.– In this expression:

– ρV_{∞}^{2} *represents twice the dynamic pressure, and* L_{c}^{2} *is the surface area or master cross-section (block surface area seen from an infinite distance upstream)* $S = L_{c}^{2}$*;*

– $\dfrac{\mu}{\rho V_{\infty} L_{c}}$ *is the inverse of the Reynolds number, but it can be written without loss of generality as* $Re = \dfrac{\rho V_{\infty} L_{c}}{\mu}$*;*

– $\dfrac{P_{\infty}}{\rho V_{\infty}^{2}}$ *is two times the pressure coefficient, the ratio of static pressure and dynamic pressure* $C_{p} = \dfrac{P_{\infty}}{\rho V_{\infty}^{2}}$.

By taking $k = \dfrac{1}{4}$, $\theta = a = -1$, we obtain:

$$F = \frac{\rho V_{\infty}^{2}}{2} L_{c}^{2}\left(\frac{\rho V_{\infty} L_{c}}{\mu}\right)\left(\frac{P_{\infty}}{\rho V_{\infty}^{2}/2}\right) = C_{p} Re \frac{\rho V_{\infty}^{2}}{2} S$$

COMMENT 2.11.– *The pressure coefficient is a dimensionless number that is used in aerodynamics and fluid dynamics to describe the change in relative pressure in the flow of a fluid. Each point of the flow has its own unique pressure coefficient.*

The pressure coefficient is a parameter used in the study of non-compressible fluids such as water, and also in low-speed flows (flows of less than one third of the speed of sound) of compressible fluids such as air. It is given by:

$$C_p = \frac{P - P_\infty}{\rho V_\infty^2 / 2}$$

Where:

– P *is the pressure at the point where the pressure coefficient is evaluated;*

– P_∞ *is the pressure in free flow (that is, away from any disturbances);*

– ρ *is the density of the fluid in free flow;*

– V_∞ *is the speed of the fluid in free flow, or the speed of the body traveling through the fluid.*

In many situations in aerodynamics and hydrodynamics, the pressure coefficient at a point close to a body is independent of the size of this body. Therefore, an engineering model can be tested in a wind tunnel or in a water tunnel.

The pressure coefficients can be determined at critical positions around the model and can be used to predict fluid pressure at these critical positions around an airplane or boat.

2.5.8. *Exercise 8: oscillation of a liquid in a U-shaped tube*

The oscillations of a non-viscous liquid in a U-shaped tube located in a vertical plane depend on the column of water L_e, the density ρ, the oscillation period T_{osc}, and the gravitational acceleration g. Determine the relationship $T_{osc} = f(L_e, \rho, g)$.

2.5.8.1. *Solution*

The oscillation period T_{osc} is written as:

$$T_{osc} = f(L_e, \rho, g) = kL_e^{\alpha} \rho^{\beta} g^{\gamma}$$

The dimension of the two parts of this relationship is written as:

$$T = L^{\alpha} (ML^{-3})^{\beta} (LT^{-2})^{\gamma}$$

This allows us to write:

$$\begin{cases} M : \beta = 0 \\ L : \alpha - 3\beta + \gamma = 0 \\ T : -2\gamma = 1 \end{cases} \Rightarrow \begin{cases} \alpha = \dfrac{1}{2} \\ \beta = 0 \\ \gamma = -\dfrac{1}{2} \end{cases}$$

The expression of the period of the oscillation of the liquid in the U-shaped tube is:

$$T_{osc} = k\sqrt{\frac{L_e}{g}}$$

The value of the proportionality constant k is determined experimentally, which is $k = \pi$, and therefore:

$$T_{osc} = 2\pi\sqrt{\frac{L_e}{g}}$$

COMMENT 2.12.– *It can also be said that the oscillation of the liquid in the U-shaped tube progresses following a sinusoidal pattern:*

$$z = z_0 \sin(\omega_0 t + \varphi)$$

with pulsation $\omega_0 = \sqrt{\frac{L_e}{g}}$*, phasing* φ *and maximum amplitude of oscillation* z_0.

The period is identical to that of a simple pendulum of length $\frac{L_e}{2}$. *In practice, these oscillations can be observed in piezometric tubes when the pressure to be measured fluctuates slowly. The effects of friction dampen the amplitude, and the time at which a column leaves its equilibrium position is all the more difficult to determine when the amplitude is smaller.*

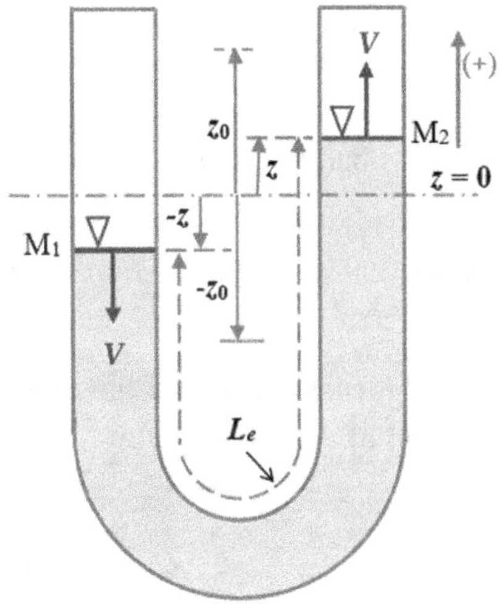

Figure 2.6. *Oscillation of a liquid in a U-shaped tube. For a color version of this figure, see www.iste.co.uk/sadchemloul/mechanics.zip*

2.5.9. *Exercise 9: a falling ball*

In a fluid, a ball of radius r traveling at a speed V is subjected to a friction force given by $F = -6\pi\mu r V$, where μ is the dynamic viscosity of the fluid.

1) What is the dimension of μ?

2) When the ball is released without any initial speed at the time $t = 0$, its speed is written as $t > 0$:

$$V(t) = a(1 - e^{-t/b})$$

where a and b are constants that depend on the characteristics of the fluid. What are the dimensions of a and b?

3) If ρ is the density of the fluid, determine the ratio $Re = \rho^{\alpha} V^{\beta} \mu^{\gamma} r^{\theta}$ where Re is a dimensionless number.

2.5.9.1. *Solutions*

1) The dimension of dynamic viscosity is:

$$[\mu] = \frac{[F]}{[V][r]} = \frac{MLT^{-2}}{LT^{-1}L} = ML^{-1}T^{-1}$$

2) The falling speed of the ball is $V = a(1-e^{-t/b})$.

The argument of exponential $-\frac{t}{b}$ is dimensionless, and thus $[b]=[t]=T$.

The term $(1-e^{-t/b})$ is dimensionless, and the dimension of the constant a is equal to that of the term of the velocity V:

$$[V]=[a]=LT^{-1}$$

3) Knowing that the dimension of the dimensionless number Re is the unit $[Re]=1=M^0L^0T^0$, we obtain:

$$[Re]=[\rho]^{\alpha}[V]^{\beta}[\mu]^{\gamma}[r]^{\theta} = (ML^{-3})^{\alpha}(LT^{-1})^{\beta}(ML^{-1}T^{-1})^{\gamma}(L)^{\theta}$$

The exponents are given by the following ratios:

$$\begin{cases} M: \alpha+\gamma=0 \\ L: -3\alpha+\beta-\gamma+\theta=0 \\ T: \beta+\gamma=0 \end{cases} \Rightarrow \begin{cases} \alpha=-\theta \\ \beta=-\theta \\ \gamma=-\theta \end{cases}$$

The dimensionless number is therefore written as:

$$Re = \left(\frac{\rho V r}{\mu}\right)^{-\theta}$$

For $\theta = -1$, we find the Reynolds number:

$$Re = \frac{\rho V r}{\mu}$$

COMMENT 2.13.– *A small sphere is dropped into a fluid, such as a viscometer with a dropped ball (see Figure 2.7). When the speed of the sphere becomes uniform, the balance of forces gives us:*

$$\overrightarrow{P_{ball}} + \overrightarrow{F_v} + \overrightarrow{F_A} = 0$$

with:

– $P_s = \frac{4}{3}\rho_{ball} g \pi r^3$ *the weight of the sphere;*

– $F_v = 4\mu\pi r V$ *the viscous force or friction between the ball and the fluid (Stokes law*[15]*);*

– $F_A = \frac{4}{3}\rho_f g \pi r^3$ *being Archimedes' buoyancy force*[16]*;*

– where ρ_s, ρ_f *are the density of the ball and the fluid, respectively, and* r *is the radius of the spherical ball.*

The balance of forces gives:

$$\frac{4}{3}\rho_f g \pi r^3 + 6\pi\mu r V - \frac{4}{3}\rho_s g \pi r^3 = 0$$

15 George Gabriel Stokes, born on August 13, 1819, in Skreen (Ireland), and who lived until February 1, 1903, was a mathematician and physicist. He made his biggest contributions in the areas of fluid mechanics, optics and geodesy. The movement of a Newtonian fluid is referred to as Navier-Stokes equations, which are the work of two scientists in the 19th Century: the mathematician and bridge engineer Henri Navier, who introduced first the notion of viscosity in the Euler equations in 1823, and the physicist George Gabriel Stokes who provided the definitive form of the equation of conservation of the quantity of movement in 1845.

16 Archimedes of Syracuse, born in Syracuse (Italy) around 287 B.C. and who lived until 212 B.C., was a physicist, mathematician and engineer.

By measuring the falling speed V, we may deduce the dynamic viscosity:

$$\mu = \frac{2}{3} r^2 g \frac{\rho_s - \rho_f}{V}$$

This ratio is valid for a Reynolds number:

$$Re = \frac{\rho_f V r}{\mu} < 5$$

The reference length for calculating the Reynolds number is the radius of the sphere.

If $Re > 5$ *swirls appear in the wake of the sphere, the Stokes formula is no longer applicable.*

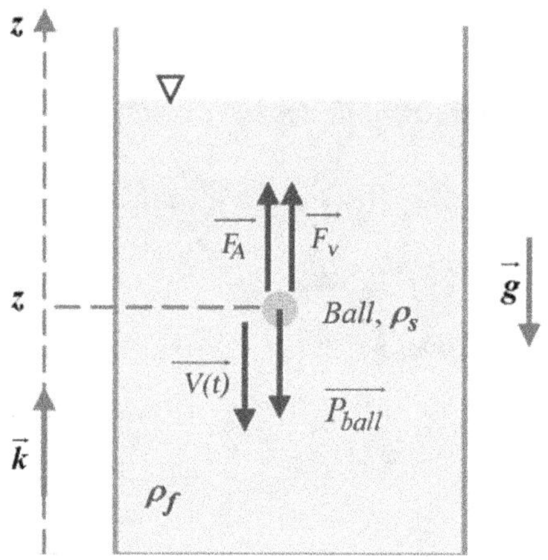

Figure 2.7. *Falling ball viscometer. For a color version of this figure, see www.iste.co.uk/sadchemloul/mechanics.zip*

COMMENT 2.14.– *The application of the basic law of dynamics gives us:*

$$\underbrace{\frac{4}{3}\rho_s \pi r^3}_{\text{ball mass}} \frac{dV}{dt} = \underbrace{\frac{4}{3}\rho_f g \pi r^3}_{\text{Archimedes' buoyancy}} + \underbrace{6\mu\pi r V}_{\text{drag force}} - \underbrace{\frac{4}{3}\rho_s \pi r^3}_{\text{ball weight}}$$

or:

$$\frac{dV}{dt}+\frac{9\mu}{2r^2\rho_s}V=g\left(1-\frac{\rho_s}{\rho_f}\right)$$

With the condition of the limits $V(t=0)=0$, *this differential equation has the solution:*

$$V(t)=g\frac{2r^2\rho_s}{9\mu}\left(1-\frac{\rho_f}{\rho_s}\right)\left(1-e^{-t/\tau_0}\right)$$

with $\tau_0=\frac{2r^2\rho_s}{9\mu}$.

The dimension of the term $g\frac{2r^2\rho_s}{9\mu}\left(1-\frac{\rho_f}{\rho_s}\right)$ *is* LT^{-1}, *and that of the term* $\tau_0=\frac{2r^2\rho_s}{9\mu}$ *is* T^{-1}.

In comparison with the ratio $V(t)=a(1-e^{-t/b})$ *of question 2), we find:*

$$a=g\frac{2r^2\rho_s}{9\mu}\left(1-\frac{\rho_f}{\rho_s}\right) \text{ and } b=\frac{2r^2\rho_s}{9\mu}$$

When $t\to\infty$, *the term with the exponent tends toward 0, and the velocity* $V(t)$ *tends toward a constant:*

$$V(t)=g\frac{2r^2\rho_s}{9\mu}\left(1-\frac{\rho_f}{\rho_s}\right)\equiv V_\infty$$

If $\rho_f<\rho_s$, *the speed* $V(t)$ *tends toward a positive value. The ball thus falls at the constant speed* V_∞, *called the "limiting speed".*

Assuming that the ball has reached its limit speed V_∞, we can use a camera to view the positions z and $z+\Delta z$ of the ball at the time t and $t+\Delta t$. On the basis of the measurements of Δz and Δt, we obtain the velocity V_∞, and thus the dynamic viscosity μ given by the formula:

$$\mu = \frac{2r^2 g\rho_s}{9V_\infty}\left(1-\frac{\rho_f}{\rho_s}\right)$$

with $\dfrac{\Delta z}{\Delta t} = V_\infty$.

2.5.10. *Exercise 10: implosion time of an air bubble*

Find the ratio giving the implosion time t_m of a water bubble that is a function of the radius r_0 of the bubble, the density ρ of the water and the resting pressure P_∞.

2.5.10.1. *Solution*

The implosion time is related to other physical quantities by the ratio:

$$t_{im} = f(r_0, \rho, P_\infty) = kr_0^\alpha \rho^\beta P_\infty^\theta$$

The dimension of this relationship is:

$$[t_{im}] = [r_0]^\alpha [\rho]^\beta [P_\infty]^\gamma \text{ or } T = (L)^\alpha (ML^{-3})^\beta (ML^{-1}T^{-2})^\gamma$$

The exponents are solutions of the following system of equations:

$$\begin{cases} M: \rho + \gamma = 0 \\ L: \alpha - 3\beta - \gamma = 0 \\ T: -2\gamma = 1 \end{cases} \Rightarrow \begin{cases} \alpha = 1 \\ \beta = 1/2 \\ \gamma = -1/2 \end{cases}$$

and therefore:

$$t_{im} = kr_0\sqrt{\frac{\rho}{P_\infty}}$$

COMMENT 2.15.– *The expression of the implosion time is established with the following assumptions:*

– we consider a gas bubble with a radius r_0 *at the time* t_0 *and of a very low internal pressure;*

– the effects of gravity on the bubble in question and the other bubbles surrounding it are negligible;

– the fluid around the single bubble is thus set in motion by the variations in the radius of the bubble;

– away from the bubble considered, the speed and pressure are such that:

$$V(r_0 = \infty, t) = 0 \text{ and } P(r_0 = \infty, t) = \infty$$

– the flow of water is supposed to be perfect, non-compressible, and homogeneous, using a Galilean reference where the bubble is immobile.

NOTE 2.3.– *This implosion occurs when the phenomenon of cavitation takes place. In the case of the propellers of a ship (see Figure 2.8a) that generate a significant depression, gas bubbles (water vapor bubbles) form. This phenomenon of cavitation is responsible for the erosion of the propeller (explosion of bubbles) and also creates a very characteristic noise made by the propeller (which makes it very harmful to ships).*

Cavitation in water is the phenomenon of the vaporization of liquid at a temperature that remains practically constant due to the effect of a drop in pressure, generating water vapor. It is therefore clearly distinct from boiling, which takes place at a virtually constant pressure, under the effect of an increase in temperature. We may distinguish the sine wave created by the air in the water. Cavitation remains centered around the edges of the blades.

These two phenomena (cavitation and boiling) are shown schematically in the thermodynamic diagram in Figure 2.8b; the function that is shown is the vapor saturation pressure of water, the equilibrium pressure between water vapor and liquid water. Cavitation manifests itself in the formation of bubbles in areas of low pressure within a flow.

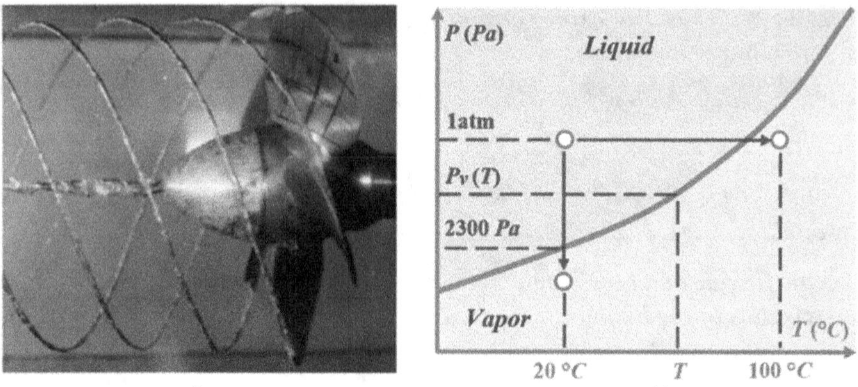

a) b)

Figure 2.8. *The phenomenon of cavitation and its thermodynamic diagram: a) an example of cavitation occurring on a propeller; b) thermodynamic diagram. For a color version of this figure, see www.iste.co.uk/sadchemloul/mechanics.zip*

2.5.11. *Exercise 11: vibration of a drop of water*

The frequency of the vibration f_v of a drop of water depends on several parameters. We will assume that surface tension is the predominant factor in the cohesion of the drop. Therefore, the physical quantities forming part of the expression of the frequency of vibration f_v are the radius of the drop r_0, the density ρ which provides the inertia, and a constant C_o, which factors into the expression of the force, and which arises due to surface tension (the dimension of C_o is that of a force per unit of length). Determine the expression of f_v.

2.5.11.1. *Solution*

The vibration frequency of a drop of water is:

$$f_v = f(r_0, \rho, C_o) = k r_0^{\alpha} \rho^{\beta} C_o^{\gamma}$$

The dimension of the constant C_o is:

$$[C_o] = [\text{surface tension}] = \frac{[\text{force}]}{[\text{lenght}]} = \frac{MLT^{-2}}{L} = MT^{-2}$$

The dimension of this relationship $f_v = f(r_o, \rho, C_o)$ is:

$$M^0 L^0 T^{-1} = L^\alpha (ML^{-3})^\beta (MT^{-2})^\gamma$$

The exponents are solutions of the following system:

$$\begin{cases} M : \beta + \gamma = 0 \\ L : \alpha - 3\beta = 0 \\ T : -2\gamma = 1 \end{cases}$$

that is:

$$\begin{cases} \alpha = -3/2 \\ \beta = -1/2 \\ \gamma = 1/2 \end{cases}$$

The vibration frequency is therefore written as[17]:

$$f_v = k \left(\frac{C_0}{r_0^3 \rho} \right)^{1/2}$$

The drop of water is an oscillator characterized by its own pulsation $f_c^2 = \dfrac{\sigma}{r_0^3 \rho}$, with σ being the surface tension of the liquid.

COMMENT 2.16.– *The cohesion of a spherical water drop is made possible by the surface tension of the drop. In this case, it will pull the drop into a spherical shape, which will therefore affect the vibration frequency of that drop. The inertia of the drop will also affect this movement, which is given by the density of the fluid.*

The size of the drop is a relevant parameter, since for a larger drop of a given fluid, the cohesion is more difficult to ensure. Therefore, there is an interplay between the surface tension and the deformation of the drop during the vibration, which is taken into account in this exercise.

17 This result had been found by Rayleigh and published in the journal *Nature* in 1915 in an article named “The Principle of Similitude”, vol. no. 95, pp. 66–68.

2.5.12. *Exercise 12: drag force of water on a ship*

Determine the resistance force (drag force) F_t exerted by water on a moving ship, which is dependent on:

– the length L_n of the ship, traveling at a speed V;

– the density ρ , dynamic viscosity μ , and surface tension σ of the water;

– the gravitational acceleration g.

2.5.12.1. *Solution*

The relationship that shows the relationship of the resistance force F_t exerted by water on a moving vessel is:

$$F_t = f(\rho, V, L_n, \mu, g, \sigma) = k\rho^{\alpha} V^{\beta} L_n^{\gamma} \mu^{\theta} g^{a} \sigma^{b}$$

The dimension of this relationship is:

$$MLT^{-2} = (ML^{-3})^{\alpha} (LT^{-1})^{\beta} (L)^{\gamma} (ML^{-1}T^{-1})^{\theta} (LT^{-2})^{a} (MT^{-2})^{b}$$

The exponents are solutions of the following system of equations:

$$\begin{cases} M : \alpha + \beta + b = 1 \\ L : -3\alpha + \beta + \gamma - \theta + a = 1 \\ T : -\beta - \theta - 2a - 2b = -2 \end{cases}$$

Since this system is incompatible, the exponents α, β, and γ are determined according to the other exponents, that is:

$$\begin{cases} \alpha = 1 - \theta - b \\ \beta = 2 - \theta - 2a - 2b \\ \gamma = 2 - \theta + a - b \end{cases}$$

The expression of the drag force is therefore:

$$F_t = k\rho^{1-\theta-b} V^{2-\theta-2a-2b} L_n^{2-\theta+a-b} \mu^{\theta} g^{a} \sigma^{b}$$

$$= k\rho V^2 L_n^2 \left(\frac{\mu}{\rho V L_n}\right)^{\theta} \left(\frac{g L_n}{V^2}\right)^{a} \left(\frac{\sigma}{\rho V^2 L_n}\right)^{b}$$

NOTE 2.4.– *The term* $\frac{\mu}{\rho V L_n}$ *is the inverse of the Reynolds number:*

$$Re = \frac{\rho V D}{\mu}$$

The term $\frac{g L_n}{V^2}$ *is the inverse of the square of Froude's number:*

$$Fr = \frac{V}{\sqrt{g L_n}}$$

The term $\frac{\sigma}{\rho V^2 L_n}$ *is Weber's number:*

$$We = \frac{\rho V^2 L_n}{\sigma}$$

The term L_n^2 *is the front surface of the ship, known as the "Master cross section".*

The water drag on the ship is written as:

$$F_t = k \rho V^2 L_n^2 F(Re, Fr, We)$$

COMMENT 2.17. – *The drag force therefore depends on three-dimensional numbers:*

– the Reynolds number "Re" *shows the effect of the forces of viscosity, or the influence of drag forces from water on the ship's hull;*

– the Froude number "Fr" *shows the effect of the forces of gravity or the influence of the wake, that is, the influence from the system of waves produced behind the vessel;*

– the Weber number "We" *shows the effect of forces of surface tension, which are negligible in this case.*

2.6. Exercises: Vaschy-Buckingham method or method of π

2.6.1. *Exercise 13: pressure drop in a pipe of circular cross-section*

Consider a fluid of density ρ and dynamic viscosity μ flowing at speed V in a horizontal pipe of diameter D. The average height of irregularities considered as roughness (absolute roughness) in this tube is ε. In considering two straight sections of the pipe at a distance of a length of L_c, determine the expression of the linear pressure drop ΔP_L between the two sections.

2.6.1.1. *Solution*

The physical quantities involved in determining the drop in pressure ΔP_L in the flow (see Figure 2.9) are:

– characteristics of the pipe: L_c, D and ε;

– the fluid's properties: ρ and μ;

– flow speed: V.

The number of physical quantities, including the pressure drop, is $n = 7$.

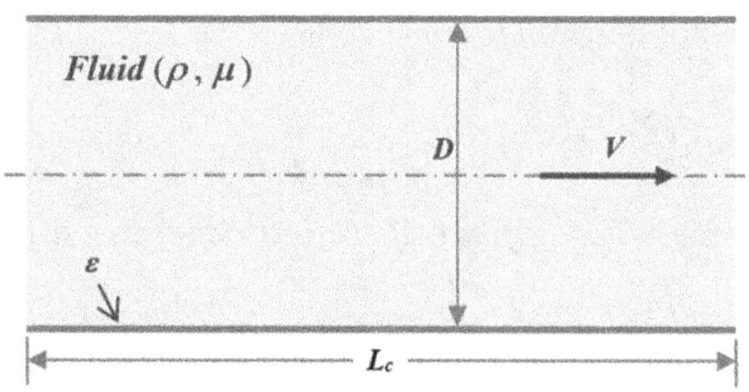

Figure 2.9. *Pressure drop in horizontal pipe of circular cross-section. For a color version of this figure, see www.iste.co.uk/sadchemloul/mechanics.zip*

The dimension of these physical quantities is:

$$[\Delta P_L] = ML^{-1}T^{-2}, \qquad [L_c] = [D] = [\varepsilon] = L, [\rho] = ML^{-3}, [\mu] = MLT^{-1},$$

$$[V] = LT^{-1}$$

The number of fundamental quantities is $k=3$.

The number of terms π involved in determining the pressure drop is $n-k=7-3=4$.

For the formation of terms in π the selected or repeated quantities are ρ, V and D, that is:

$$\begin{cases}\pi_1=\rho^{\alpha}V^{\beta}D^{\gamma}L_c\\ \pi_2=\rho^{\alpha}V^{\beta}D^{\gamma}\varepsilon\\ \pi_3=\rho^{\alpha}V^{\beta}D^{\gamma}\mu\\ \pi_4=\rho^{\alpha}V^{\beta}D^{\gamma}\Delta P_L\end{cases}$$

We will consider the dimensions of terms in π:

$$[\pi_1]=M^0L^0T^0=(ML^{-3})^{\alpha}(LT^{-1})^{\beta}(L)^{\gamma}L$$

$$\begin{cases}M:\alpha=0\\ L:-3\alpha+\beta+\gamma+1=0\\ T:-\beta=0\end{cases}\Rightarrow\begin{cases}\alpha=-1\\ \beta=0\\ \gamma=-1\end{cases}$$

The term $\pi_1=\dfrac{L_c}{D}$ is the ratio between the length and diameter of the pipe.

$$[\pi_2]=M^0L^0T^0=(ML^{-3})^{\alpha}(LT^{-1})^{\beta}(L)^{\gamma}L$$

$$\begin{cases}M:\alpha=0\\ L:-3\alpha+\beta+\gamma+1=0\\ T:-\beta=0\end{cases}\Rightarrow\begin{cases}\alpha=0\\ \beta=0\\ \gamma=-1\end{cases}$$

The term $\pi_2=\dfrac{\varepsilon}{D}$ is the relative roughness: the ratio of the absolute roughness of the inner surface of the pipe to the diameter of the pipe.

$$[\pi_3]=ML^{-1}T^{-1}=(ML^{-3})^{\alpha}(LT^{1})^{\beta}(L)^{\gamma}ML^{-1}T^{-1}$$

$$\begin{cases} M: \alpha + 1 = 0 \\ L: -3\alpha + \beta + \gamma + 1 = 0 \\ T: -\beta - 1 = 0 \end{cases} \Rightarrow \begin{cases} \alpha = -1 \\ \beta = -1 \\ \gamma = -1 \end{cases}$$

The term $\pi_3 = \dfrac{\mu}{\rho V D}$ represents the inverse of the Reynolds number; a dimensional analysis allows us to write $\pi_3 = Re = \dfrac{\rho V D}{\mu}$.

$$[\pi_4] = M^0 L^0 T^0 = (ML^{-3})^{\alpha} (LT^{-1})^{\beta} (L)^{\gamma} ML^{-1}T^{-2}$$

$$\begin{cases} M: \alpha + 1 = 0 \\ L: -3\alpha + \beta + \gamma + 1 = 0 \\ T: -\beta - 2 = 0 \end{cases} \Rightarrow \begin{cases} \alpha = -1 \\ \beta = -2 \\ \gamma = 0 \end{cases}$$

The term $\pi_4 = \dfrac{\Delta P_L}{\rho V^2}$ is the relationship between the pressure drop and the dynamic pressure, which is the pressure coefficient.

The terms in π thus verify the expression:

$$F(\pi_1, \pi_2, \pi_3, \pi_4) = F(\frac{L_c}{D}, \frac{\varepsilon}{D}, Re, \frac{\Delta P_L}{\rho V^2}) = 0$$

The expression of the loss of pressure is thus given by:

$$\Delta P_L = \rho V^2 F_1(\frac{L_c}{D}, \frac{\varepsilon}{D}, Re)$$

NOTE 2.5.– *As the relationship* $\dfrac{L_c}{D}$ *is constant for all straight pipes with circular cross-sections, the pressure drop is therefore written:*

$$\Delta P_L = \rho V^2 \frac{L_c}{D} F_2(\frac{\varepsilon}{D}, Re) = \lambda \rho V^2 \frac{L_c}{D}$$

The function $\lambda = F_2(\frac{\varepsilon}{D}, Re)$ *is called the "linear pressure drop coefficient". This coefficient depends on the flow regime (Reynolds number Re) and the relative roughness* $\frac{\varepsilon}{D}$. *The determination of* λ *is created from the relationships in Table 2.3 or from the Moody diagram in Figure 2.5 of exercise 6.*

2.6.2. *Exercise 14: friction forces on a flat plate*

Determine the relationship giving the friction (or drag) force F_t exerted by a fluid moving at a speed of V on a flat plate of length L_p and width l_p. The flow occurs parallel to the length of the plate. The fluid has a density of ρ and a dynamic viscosity of μ .

2.6.2.1. *Solution*

The physical quantities are: the friction force F_t, the velocity of the fluid V, the plate length L_p and width l_p, the density ρ and the dynamic viscosity μ of the fluid, that is, $n = 6$.

The dimensions of these physical quantities are:

$$[F_t] = MLT^{-2},\ [V] = LT^{-1},\ \left[L_p\right] = [l_P] = L,\ [\rho] = ML^{-3},\ [\mu] = ML^{-1}T^{-1}$$

The number of fundamental quantities is $k = 3$.

The number of terms in π is $n - k = 6 - 3 = 3$. For the formation of terms in π the quantities chosen are ρ , V and L_p:

$$\begin{cases} \pi_1 = \rho^{\alpha} V^{\beta} L_p^{\gamma} F_t \\ \pi_2 = \rho^{\alpha} V^{\beta} L_p^{\gamma} l_p \\ \pi_3 = \rho^{\alpha} V^{\beta} L_p^{\gamma} \mu \end{cases}$$

The dimension of the term π_1 gives us:

$$M^0 L^0 T^0 = (ML^{-3})^{\alpha} (LT^{-1})^{\beta} (L)^{\gamma} MLT^{-2}$$

$$\begin{cases} M : \alpha + 1 = 0 \\ L : -3\alpha + \beta + \gamma + 1 = 0 \\ T : -\beta - 2 = 0 \end{cases} \Rightarrow \begin{cases} \alpha = -1 \\ \beta = -2 \\ \gamma = -2 \end{cases}$$

The expression of the term π_1 is $\pi_1 = \dfrac{F_t}{\rho V^2 L_p^2}$.

The dimension of the term π_2 gives us:

$$M^0 L^0 T^0 = (ML^{-3})^{\alpha} (LT^{-1})^{\beta} (L)^{\gamma} L$$

$$\begin{cases} M : \alpha = 0 \\ L : -3\alpha + \beta + \gamma + 1 = 0 \\ T : -\beta = 0 \end{cases} \Rightarrow \begin{cases} \alpha = 0 \\ \beta = 0 \\ \gamma = -1 \end{cases}$$

The expression of the term π_2 is $\pi_2 = \dfrac{l_p}{L_p}$, which is the ratio of the width to length of the flat plate.

The dimension of the term π_3 gives us:

$$M^0 L^0 T^0 = (ML^{-3})^{\alpha} (LT^{-1})^{\beta} (L)^{\gamma} ML^{-1}T^{-1}$$

$$\begin{cases} M : \alpha + 1 = 0 \\ L : -3\alpha + \beta + \gamma + 1 = 0 \\ T : -\beta - 1 = 0 \end{cases} \Rightarrow \begin{cases} \alpha = -1 \\ \beta = -1 \\ \gamma = -1 \end{cases}$$

The expression of the term π_3 is $\pi_3 = \dfrac{\mu}{\rho V L_p} = \dfrac{1}{Re}$, which is the inverse of the Reynolds number Re.

Dimensional analysis makes it possible to write the following: $\pi_3 = Re = \dfrac{\rho V L_p}{\mu}$

The terms in π thus verify the expression:

$$F(\pi_1, \pi_2, \pi_3) = F(\frac{F_t}{\rho V^2 L_p^2}, Re, \frac{l_p}{L_p}) = 0$$

The expression of the frictional force is $F_t = \rho V^2 L_p^2 F_1(Re, \dfrac{l_p}{L_p})$.

COMMENT 2.18.– *The ratio* $\dfrac{l_p}{L_p}$ *is constant for all flat plates (with similar geometrical forms).*

The force of friction on both sides of the plate is written as:

$$F_t = \rho V^2 L_p l_p F_2(Re)$$

In this relationship, the product $L_p l_p$ *represents the area of a single face of the plate* ($S_{plate} = L_p l_p$).

For one side of the plate $F_t = \rho \dfrac{V^2}{2} L_p l_p F_2(Re)$.

The relationship $\dfrac{F_t}{S_{plate}} = \tau_{plate}$ *is the stress along the wall and the relationship* $\dfrac{\tau_p}{\rho V^2 / 2} = C_F = F_2(Re)$ *is the coefficient of friction (or drag coefficient) that is dependent on the Reynolds number. The term* $\rho V^2/2$ *represents the dynamic pressure, and thus the coefficient of friction (or drag coefficient) is the relationship between the shear stress along the wall and the dynamic pressure.*

The stress measured on the wall depends on the position x *taken,* $\tau_p(x)$, *on the length of the plate (direction of the flow), and thus we may refer to the local coefficient of friction:*

$$C_F(x) = \frac{\tau_p(x)}{\rho V^2/2} \quad \textit{with: } \tau_p(x) = \mu \left.\frac{\partial u(x,y)}{\partial y}\right|_{y=0}$$

with $u(x, y)$ *being the speed profile inside the boundary layer.*

Upstream of the flat plate, the fluid has an infinite upstream velocity of U_∞. *The flow of the fluid is parallel to the plate, the thickness of which is generally considered negligible along its leading edge (see figure 2.10). Outside the boundary layer, the velocity of the fluid remains constant at the value of* U_∞, *which is the region of the free stream where the inertial forces are very large with respect to the viscous forces. Inside the boundary layer, the viscous forces are greater than the inertial forces, and the velocity gradient is not zero.*

The thickness $\delta(x)$ *of the boundary layer of* x *is defined as:*

$$u(x, y = \rho) / U_\infty = 0.99$$

In the case of the laminar boundary layer, the momentum transfer is ensured by the viscous friction within the entire boundary layer. The local coefficient of friction varies as given by $Re_x^{0,5}$ (Re_x *is the Reynolds number determined along the axis of* x *), and the local tangential stress decreases as we move away from the leading edge of the plate.*

In the turbulent boundary layer, the transport of the quantity of movement is only provided by the viscous friction within a very thin sub-layer, known as the laminar boundary layer. Above this layer, the transport of the amount of movement is provided by turbulent lateral fluctuations. The local coefficient of friction is essentially independent of the position x *taken on the plate, and the drag coefficient is nearly constant. This result shows that the drag coefficient is independent of the Reynolds number in the turbulence scheme, which means that the drag force varies at the level of* U_∞^2.

In the general case, the velocity at infinity U_∞ *is different from the speed* U_e *outside the boundary layer of turbulence.*

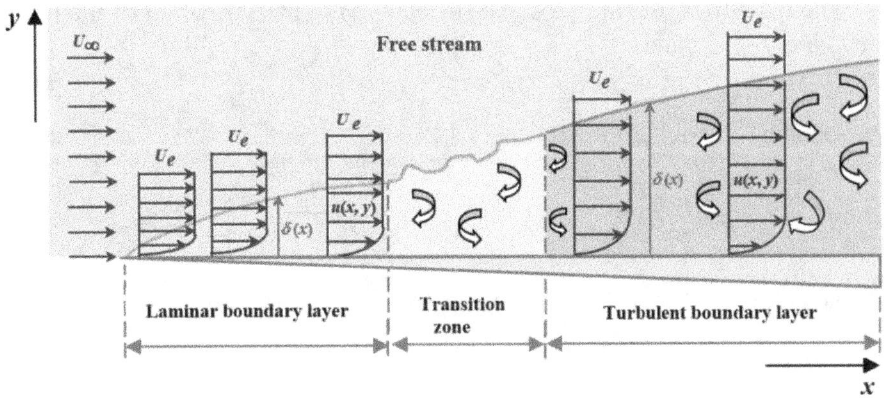

Figure 2.10. *Flow near a flat wall (dynamic boundary layers). For a color version of this figure, see www.iste.co.uk/sadchemloul/mechanics.zip*

2.6.3. *Exercise 15: drag force exerted on a sphere*

Determine the relationship of the drag force F_t exercised by the uniform flow of a non-compressible viscous fluid over a sphere. This drag depends on the density ρ and dynamic viscosity μ of the fluid, the speed and pressure at infinity of the flow, V_∞ and P_∞, and the diameter of the sphere.

2.6.3.1. *Solution*

The drag F_t exerted on an object is expressed as:

$$F_t = f(V_\infty, P_\infty, D, \rho, \mu)$$

The number of physical quantities is therefore $n = 6$.

The dimension of these physical quantities is:

$$[F_t] = MLT^{-2},\ [V_\infty] = LT^{-1},\ [P_\infty] = ML^{-1}T^{-2},\ [D] = L,\ [\rho] = ML^{-3},$$
$$[\mu] = ML^{-1}T^{-1}$$

The number of fundamental quantities is $k=3$. Thus, there are three ($n-k=3$) terms in π.

The quantities that are repeated (or chosen) in the formation of the terms in π are ρ, V_∞, and D:

$$\begin{cases} \pi_1 = \rho^\alpha V_\infty^\beta D^\gamma F_t \\ \pi_2 = \rho^\alpha V_\infty^\beta D^\gamma \mu \\ \pi_3 = \rho^\alpha V_\infty^\beta D^\gamma P_\infty \end{cases}$$

The dimension of the term π_1 gives us:

$$M^0 L^0 T^0 = (ML^{-3})^\alpha (LT^{-1})^\beta (L)^\gamma MLT^{-2}$$

$$\begin{cases} \alpha+1=0 \\ -3\alpha+\beta+\gamma+1=0 \\ -\beta-2=0 \end{cases} \Rightarrow \begin{cases} \alpha=-1 \\ \beta=-2 \\ \gamma=-2 \end{cases}$$

The expression of the term π_1 is $\pi_1 = \dfrac{F_t}{\rho V_\infty^2 D^2}$.

The dimension of the term π_2 gives us:

$$M^0 L^0 T^0 = (ML^{-3})^\alpha (LT^{-1})^\beta (L)^\gamma ML^{-1}T^{-1}$$

$$\begin{cases} M: \alpha+1=0 \\ L: -3\alpha+\beta+\gamma+1=0 \\ T: -\beta-1=0 \end{cases} \Rightarrow \begin{cases} \alpha=-1 \\ \beta=-1 \\ \gamma=-1 \end{cases}$$

The expression of the term π_2 is $\pi_2 = \dfrac{\mu}{\rho V_\infty D}$. This is the inverse of the Reynolds number, but we can write it as:

$$\pi_2 = Re = \frac{\rho V_\infty D}{\mu}$$

This is allowed in dimensional analysis.

The dimension of the term π_3 gives us:

$$M^0 L^0 T^0 = (ML^{-3})^{\alpha}(LT^{-1})^{\beta}(L)^{\gamma} ML^{-1}T^{-2}$$

$$\begin{cases} M: \alpha + 1 = 0 \\ L: -3\alpha + \beta + \gamma + 1 = 0 \\ T: -\beta - 2 = 0 \end{cases} \Rightarrow \begin{cases} \alpha = -1 \\ \beta = -2 \\ \gamma = 0 \end{cases}$$

The expression of the term π_3 is $\pi_3 = \dfrac{P_\infty}{\rho V_\infty^2 D}$.

The three terms in π thus verify the expression:

$$F\left(\frac{F_t}{\rho V_\infty^2 D^2}, Re, \frac{P_\infty}{\rho V_\infty D}\right) = 0$$

The drag is given by the relationship:

$$F_t = \rho V_\infty^2 D^2 F_1\left(\mathrm{Re}, \frac{P_\infty}{\rho V_\infty D}\right)$$

NOTE 2.6.– *The pressure P_∞ has an effect on determining the momentum balance through its gradient; this field is defined to within a constant. Therefore, any translation of the value P_∞ leaves the drag force F_t as fixed. It then follows that F_t is independent of the non-dimensional quantity $P_\infty / \rho V_\infty^2$. The drag can only be expressed as a function of the only dimensional number, which is the Reynolds number Re, giving us:*

$$F_t = \rho V_\infty^2 D^2 F(Re)$$

In practice, the drag force is given in the following form:

$$F_t = C_D S(\rho V_\infty^2 / 2)$$

with C_D *being the drag coefficient, dependent on the Reynolds number,* $C_D = F(Re)$, $S = \pi D^2 / 4$ *being the apparent surface area of the sphere in the direction of the flow, and* $\rho V_\infty^2 / 2$ *being the dynamic pressure. This empirical result is supported by analytical and experimental studies. In effect, an analytical development based on the theory of potential flows shows us that for flows with very low Reynolds numbers* ($Re \approx 1$)*, the drag coefficient is expressed analytically in the following way:*

$$C_D = 24 / Re$$

On the other hand, experimental work shows that for the larger Reynolds numbers, the variation of C_D *can be shown by what is presented in Figure 2.11. This figure shows that for a Reynolds number* $10^3 < Re < 2\times10^5$, *the coefficient* C_D *maintains a largely constant value, independently of* Re, *and around the value 0.44.*

For a Reynolds number of the order of $Re = 3\times10^5$, *the value* C_D *drops sharply. This fall is due to the drag crises phenomenon.*

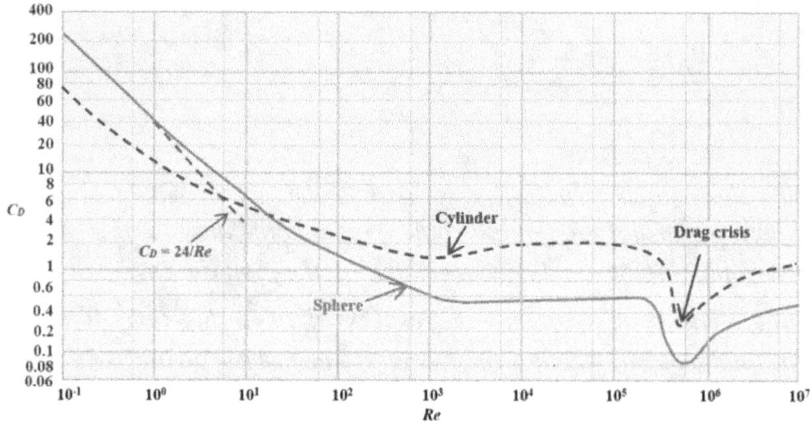

Figure 2.11. *Variation of the drag coefficient as a function of the Reynolds number in the case of a sphere and a cylinder. For a color version of this figure, see www.iste.co.uk/sadchemloul/mechanics.zip*

CONCLUSION.– *The dimensional analysis for this study scenario has allowed us to simplify the study considerably. Effectively, it showed that the drag depends on a single, non-dimensional parameter: the Reynolds number. More in-depth studies (analytical or experimental studies) allow us to better explain such a relationship.*

For commonly used non-spherical obstacles, Table 2.5 gives the drag coefficient values measured experimentally for some commonly used two-dimensional and three-dimensional objects, and for a range of Reynolds numbers greater than 10^4.

	Shape	C_D		Shape	C_D
Two dimensional	Square			Half tube	
	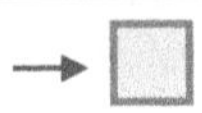	2.1			1.2
	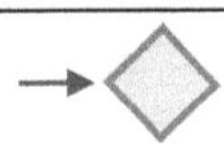	1.6		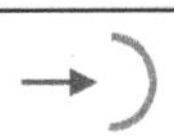	2.3
	Half cylinder			Equilateral triangle	
	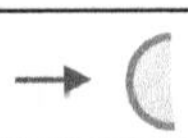	1.2			1.6
	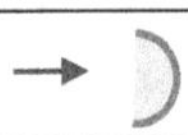	1.7		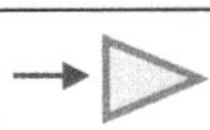	2.0
Three dimensional	Cub			Spherical cap	
	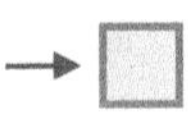	1.07		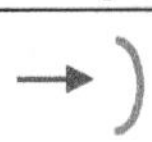	0.4
	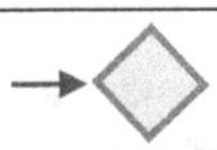	0.81		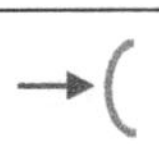	1.4
	Cone	θ(°)	C_D	Parachute	
	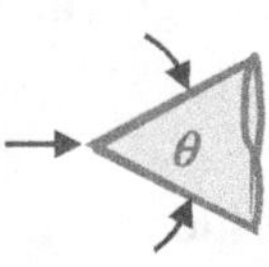	10	0.30		1.2
		20	0.40		
		30	0.55		
		40	0.65		
		60	0.80		
		90	1.15		

Table 2.5. *Drag coefficient values for two-dimensional and three-dimensional objects calculated on the front surface area for a Reynolds number greater than or equal to 10^5. For a color version of this figure, see www.iste.co.uk/sadchemloul/mechanics.zip*

2.6.4. *Exercise 16: hydraulic jump*

In a channel with a horizontal bottom, a study is to be conducted of linear geometric characteristic X of a jump in the hydraulic flow formed by the liquid flowing under a canal gate with height a subjected to a load H, with the runoff downstream having the value of y. Conduct a dimensional study of the problem, neglecting the viscosity forces and the surface tension.

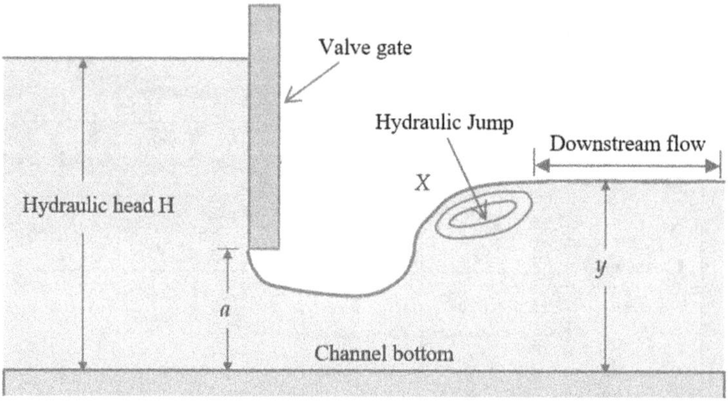

Figure 2.12. *Flow in a channel. For a color version of this figure, see www.iste.co.uk/sadchemloul/mechanics.zip*

2.6.4.1. *Solution*

REVIEW.– In flows on open surfaces (see Figure 2.12), the effects of viscosity are often negligible. In addition, the effects of surface tension are too small in comparison with the effects of gravity.

Thus, the factors having an effect are: the load H, the lifting of the gate a, the water flow y downstream of the gate, the gravitational acceleration g, and the density ρ.

The linear geometric characteristic X is written as:

$$X = f(H, a, y, g, \rho)$$

The number of physical quantities is therefore $n = 6$.

The dimension of these physical quantities is:

$$[X]=[H]=[a]=[y]=L,\ [g]=LT^{-2},\ [\rho]=ML^{-3}$$

The number of fundamental quantities is $k=3$.

The number of terms π involved in determining the pressure drop is $n-k=3$.

For the formation of terms in π the quantities chosen are ρ, g, and H, which gives us:

$$\begin{cases}\pi_1=\rho^{\alpha}g^{\beta}H^{\gamma}X\\ \pi_2=\rho^{\alpha}g^{\beta}H^{\gamma}a\\ \pi_3=\rho^{\alpha}g^{\beta}H^{\gamma}y\end{cases}$$

NOTE 2.7.– *This system is formed from the same equation, because X, a and y have the same dimension.*

The dimension of the three terms in π is:

$$(ML^{-3})^{\rho}(LT^{-1})^{\beta}(L)^{\gamma}L$$

$$\begin{cases}M:\alpha=0\\ L:-3\alpha+\beta+\gamma+1=0\\ T:-\beta=0\end{cases}\Rightarrow\begin{cases}\alpha=0\\ \beta=0\\ \gamma=-1\end{cases}$$

$$\pi_1=\frac{X}{H},\ \pi_2=\frac{a}{H},\ \pi_3=\frac{y}{H}$$

These three terms in π verify the relationship:

$$F(\pi_1,\pi_2,\pi_3)=0 \text{ or } X=HF_1(\frac{a}{H},\frac{y}{H})$$

COMMENT 2.19.– *It should be noted that the Froude number does not have any effect when it is a flow on an open surface where the inertial forces and gravitational forces have a great influence. In reality, the relationships of*

$\frac{H}{a}$ and $\frac{H}{y}$ are equivalent to Froude numbers, because the speed arising from the load H on the valve is proportional to $\sqrt{2gH}$ *(the Torricelli ratio).*

The result obtained shows that the physical quantities ρ *and* g *have no direct influence on the physical phenomenon of the hydraulic canal gates. The gate is the main means used by hydraulic structures to dissipate energy. It is formed during the abrupt transition from a torrential flow (or super-critical flow) to streaming flow (or sub-critical flow). During this transition, a stationary wave is formed and the energy is then dissipated through turbulence.*

Thus, one of the important roles of these works is to convert the flow of the stream (usually a runoff flow) to a torrential flow, so that the gate of the water jump can be formed. This is achieved either through a flow on an inclined slope higher than the critical slope (inclined fall), or through the free fall of a body of water (vertical fall). In order to attain the correct size of these hydraulic works, it is essential to have correct knowledge of the characteristics of the water drops. This primarily includes the water levels upstream and downstream of the drop-off, and the length necessary for the completion of this drop.

2.6.5. *Exercise 17: flow through a thin-walled spillway with a horizontal crested*

Using dimensional analysis, find the formula giving the flow Q_v of a spillway with thin walls and a horizontal limiter. This flow depends on the width of the spillway *B*, the upstream height of the water *H*, and the gravitational acceleration *g*. We will not consider viscosity and surface tension.

2.6.5.1. *Solution*

The effects of viscosity and surface tension are negligible, only inertia and gravity forces will be considered, and the volumetric mass ρ of the liquid will be eliminated. The relationship of the flow rate is written as $Q_v = f(B, H, g)$, and the number of physical quantities is $n = 4$.

The dimension of the physical quantities is:

$$[Q_v] = L^3T^{-1},\ [B] = [H] = L,\ [G] = LT^{-2}$$

The number of fundamental quantities is $k = 2$, that is, two terms in π:

$$\begin{cases} \pi_1 = g^{\alpha} H^{\beta} B \\ \pi_2 = g^{\alpha} H^{\beta} Q_v \end{cases}$$

We will consider the dimensions of π terms in:

$$\begin{cases} L: 3\alpha + \beta + 1 = 0 \\ T: 2\alpha = 0 \end{cases} \Rightarrow \begin{cases} \alpha = 0 \\ \beta = -1 \end{cases}$$

and therefore $\pi_1 = \dfrac{B}{H}$.

$$\begin{cases} L: \alpha + \beta + 3 = 0 \\ T: -2\alpha - 1 = 0 \end{cases} \Rightarrow \begin{cases} \alpha = -1/2 \\ \beta = -5/2 \end{cases}$$

and therefore $\pi_2 = \dfrac{Q_v}{g^{1/2} H^{5/2}}$.

Therefore, we can write:

$$F\left(\frac{Q_v}{g^{1/2} H^{5/2}}, \frac{B}{H}\right) = 0 \text{ or } \frac{Q_v}{g^{1/2} H^{5/2}} = F_1\left(\frac{B}{H}\right)$$

The expression of the flow rate Q_v of a thin-walled spillway with a horizontal limiter is:

$$Q_v = F_1\left(\frac{B}{H}\right) g^{1/2} H^{5/2} \qquad [2.12]$$

COMMENT 2.20.– *We can solve this problem by considering the volume flow per unit width* Q_v / B, *which depends on the height of the water H above the limiter and the gravitational acceleration g.*

The relationship of the flow is written as:

$$\frac{Q_v}{H} = f(H, g)$$

The number of physical quantities to take into account is $n = 3$.

The dimension of physical quantities is:

$$[H] = L, \left[\frac{Q_v}{B}\right] = L^2T^{-1}, [g] = LT^{-2}$$

The number of fundamental quantities is $k = 2$. *One can thus form a single non-dimensional product which is equal to a constant K:*

$$\pi_1 = g^{\alpha} H^{\beta} \frac{Q_v}{B} = K$$

The dimension of this term gives us:

$$\begin{cases} L : \alpha + \beta + 2 = 0 \\ T : 2\alpha - 1 = 0 \end{cases} \Rightarrow \begin{cases} \alpha = -1/2 \\ \beta = -3/2 \end{cases}$$

$$\pi_1 = \frac{Q_v}{g^{1/2} H^{3/2} B} = K$$

or:

$$Q_v = KBg^{1/2}H^{3/2} \qquad [2.13]$$

The constant K *has no dimension and will be determined from experience or from considerations outside of dimensional analysis.*

When considering similar spillways geometrically, the function $F_1(\frac{B}{H})$ *of the relationship [2.12] is linear, that is* $F_1 = K\frac{B}{H}$.

The relationship [2.12] is thus written as:

$$Q_v = K\frac{B}{H}g^{1/2}H^{5/2} = KBg^{1/2}H^{3/2}$$

We obtain the relationship [2.13].

Weirs are devices used to measure and control runoff flow rates. They have other uses than for measuring flows. In rural areas, weir barriers are used for water management and floodwater rerouting.

Protective weirs allow for excess flow levels to be redirected when dams are in danger.

Weirs consist of a wall, which can be either thin or thick, placed vertically and left open along the top. This opening is often rectangular, triangular, as we will examine in the next exercise, trapezoidal, or parabolic.

2.6.6. *Exercise 18: flow through a triangular weir*

Using dimensional analysis, find the formula giving the flow Q_v through a triangular spillway with thin walls as a function of the water height *H*, a spillway width B and a gravitational acceleration g. We will not consider the viscosity of the fluid and the surface tension.

2.6.6.1. *Solution*

The number of physical quantities is therefore $n = 4$.

The dimension of physical quantities is:

$$[Q_v] = LT^{-3},\ [B] = [H] = L,\ [g] = LT^{-2}$$

The number of fundamental quantities is $k = 2$. Thus, there are two terms in π.

If g and H are taken as chosen or repeated quantities, the terms π are written as:

$$\begin{cases} \pi_1 = g^{\alpha} H^{\beta} B \\ \pi_2 = g^{\alpha} H^{\beta} Q_v \end{cases}$$

The dimension of the terms π gives us:

$$\begin{cases} L: \alpha + \beta + 1 = 0 \\ T: -2\alpha = 0 \end{cases} \Rightarrow \begin{cases} \alpha = 0 \\ \beta = -1 \end{cases}$$

and therefore $\pi_1 = \dfrac{B}{H}$.

$$\begin{cases} L: \alpha + \beta + 3 = 0 \\ T: -2\alpha - 1 = 0 \end{cases} \Rightarrow \begin{cases} \alpha = -1/2 \\ \beta = -5/2 \end{cases}$$

and therefore $\pi_2 = \dfrac{Q_v}{g^{1/2} H^{5/2}}$.

$$F(\pi_1, \pi_2) = F\left(\frac{B}{H}, \frac{Q_v}{g^{1/2} H^{5/2}}\right) = 0$$

The expression of the flow is $Q_v = \dfrac{B}{H} g^{1/2} H^{5/2}$.

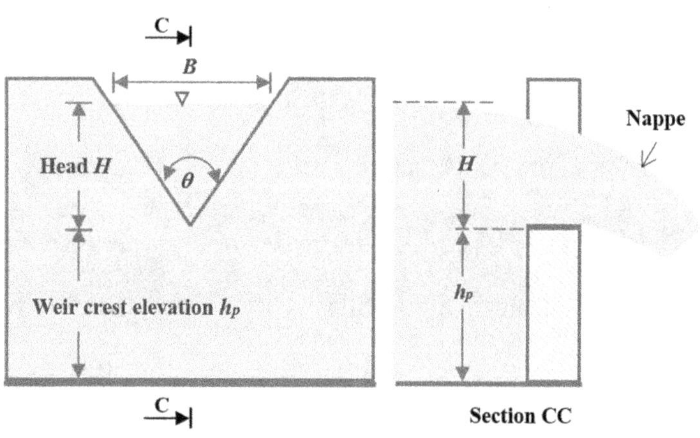

Figure 2.13. *Dimensions of a triangular spillway. For a color version of this figure, see www.iste.co.uk/sadchemloul/mechanics.zip*

COMMENT 2.21.– *If the effects of viscosity and surface tension are negligible, only inertia and gravity forces will be considered, and the volumetric mass of the liquid will be eliminated.*

The height H is counted positively above the top of the spillway. We can also find the following solution:

$$\begin{cases} \pi_1 = \dfrac{Q_v}{g^{1/2} H^{5/2}} \\ \pi_2 = \theta \end{cases}$$

$$F(\pi_1, \pi_2) = 0 \ \ or \ \ F(\frac{Q_v}{g^{1/2} H^{5/2}}, \theta) = 0$$

The expression of the flow is:

$$Q_v = F_1(\theta) g^{1/2} H^{5/2}$$

The function $F_1(\theta)$ *has no dimension and will be determined from experience or from considerations outside of dimensional analysis. This function is proportional to that obtained in the first solution:*

$$F_1(\frac{B}{H})$$

because:

$$tg\frac{\theta}{2} = \frac{B}{2H}$$

The general flow formula for a thin-walled triangular spillway is:

$$Q_v = \frac{8}{15} C_d tg(\frac{\theta}{2}) H^2 \sqrt{2gH} = \frac{8}{15} \sqrt{2g} C_d tg(\frac{\theta}{2}) H^{5/2}$$

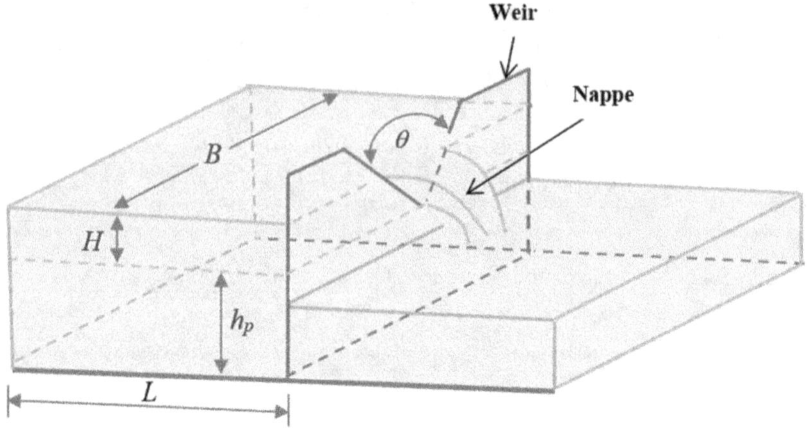

Figure 2.14. *Triangular notch weir with thin wall, placed in a hydraulic channel. For a color version of this figure, see www.iste.co.uk/sadchemloul/mechanics.zip*

C_d is the flow coefficient that depends on the type of spillway (shape and size). It represents all parameters (see Figure 2.14) that influence the flow, including:

– the angle of the opening θ or angle at the peak;

– the height of the overflow wall h_p, also called the "scoop height";

– the water load H above the edge of the wall, measured before the sagging of the surface occurs once the flow arrives;

– the flow supply width of channel B;

– the total length of the flow supply of channel L.

2.6.7. *Exercise 19: volume of a bubble*

Considering that the volume v of a bubble formed slowly within a liquid out of an orifice with a fairly small diameter only depends on the volumetric weight ϖ of the liquid, the liquid's surface tension σ, and the dynamic viscosity μ. Under these conditions, give the expression of this volume.

2.6.7.1. *Solution*

The number of physical quantities is therefore $n = 4$.

The dimensions of these physical quantities are:

$$[v],\ [\varpi] = ML^{-2}T^{-2},\ [\sigma] = MT^{-2}, [\mu] = ML^{-1}T^{-1}$$

The number of fundamental quantities is $k = 3$.

So there is only one term in π, which is $n - k = 4 - 3 = 1$.

If ϖ , σ, and μ are taken as chosen or repeated quantities, the term π is written as $\pi = \varpi^{\alpha}\sigma^{\beta}\mu^{\gamma}v$.

The dimension of this term gives us:

$$M^0L^0T^0 = (ML^{-2}T^{-2})^{\alpha}(MT^{-2})^{\beta}(ML^{-1}T^{-1})^{\gamma}L^3$$

$$\begin{cases} M: \alpha + \beta + \gamma = 0 \\ L: 2\alpha - \gamma + 3 = 0 \\ T: 2\alpha + 2\beta + \gamma = 0 \end{cases} \Rightarrow \begin{cases} \alpha = 3/2 \\ \beta = -3/2 \\ \gamma = 0 \end{cases}$$

The term in π is written as $\pi = K = \left(\dfrac{\varpi}{\sigma}\right)^{3/2} v$ with K being constant. The expression of the volume of the bubble is $v = K\left(\dfrac{\sigma}{\varpi}\right)^{3/2}$

NOTE 2.8.– *The same solution can be obtained by assuming that the forming of the bubble is done very slowly, and therefore the effect of dynamic viscosity is negligible. In this case, the problem becomes almost static, and it is advisable not to introduce time units.*

By choosing force as the fundamental unit, the three variables v , ϖ, *and* σ *can only be expressed according to the two fundamental units of force and length:*

$$[v] = L^3,\ [\varpi] = FL^{-3},\ [\sigma] = FL^{-1}$$

Thus, there is only one term in π, *which is:*

$$\pi = \frac{\varpi v^{2/3}}{\sigma} = K$$

And thus, the relationship can be written as:

$$v = K_1 \left(\frac{\sigma}{\varpi} \right)^{3/2}$$

with $K_1 = (K)^{3/2}$ *as a constant.*

The forming of the bubble created slowly in a liquid through a fairly small diameter orifice depends on the following:

– the geometric criterion: if $\frac{L}{d^4} > 10^{12}\, m^{-3}$, *the formation of bubbles is made through a constant flow, with d and L being the diameter and length of the orifice, respectively;*

– the criterion based on the drop in pressure ΔP *between the inside and outside of the bubble. It is necessary to have* $\frac{\Delta P}{(4\sigma / d)} > 1$ *in order for the bubble to form.*

The volume of the bubbles increases with the flow of the liquid within which the bubbles form.

2.6.8. *Exercise 20: flow through an orifice*

A container holds a liquid of density ρ at a height h and viscosity μ, above which it will have the pressure P_o. Through an orifice with diametered and sharp edges, a volume flow Q_v flows toward the outside, which is assumed at zero pressure.

Assuming a size h before d, what form of the law would give Q_v ?

2.6.8.1. *Solution*

The number of physical quantities is therefore $n = 7$.

$$Q_v = f(\rho, \mu, d, g, h, P_o)$$

The dimension of the physical quantities is:

$$[Q_v] = L^3T^{-1},\ [\rho] = ML^{-3},\ [d] = [h] = L,\ [g] = LT^{-2},\ [P_o] = ML^{-1}T^{-2}$$

The number of fundamental quantities is $k = 3$, that is, four terms in π:

Or ρ, d, and P_o, the quantities repeated or chosen.

The terms in π that are formed are:

$$\begin{cases} \pi_1 = \rho^\alpha d^\beta P_o^\gamma Q_v \\ \pi_2 = \rho^\alpha d^\beta P_o^\gamma h \\ \pi_3 = \rho^\alpha d^\beta P_o^\gamma g \\ \pi_4 = \rho^\alpha d^\beta P_o^\gamma \mu \end{cases}$$

The dimension of the term π_1 gives us:

$$\begin{cases} M: \alpha + \gamma = 0 \\ L: -3\alpha + \beta - \gamma + 3 \Rightarrow \\ T: -2\gamma - 1 = 0 \end{cases} \begin{cases} \alpha = 1/2 \\ \beta = -2 \\ \gamma = -1/2 \end{cases}$$

The term π_1 is written as $\pi_1 = \dfrac{Q_v}{d^2\sqrt{P_o/\rho}}$.

The dimension of the term π_2 gives us:

$$\begin{cases} M: \alpha + \gamma = 0 \\ L: -3\alpha + \beta - \gamma + 1 = 0 \Rightarrow \\ T: -2\gamma = 0 \end{cases} \begin{cases} \alpha = 0 \\ \beta = -1 \\ \gamma = 0 \end{cases}$$

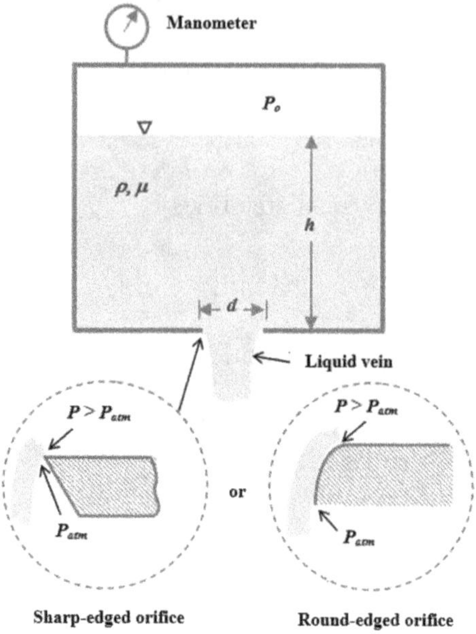

Figure 2.15. *Determining the volumetric flow through an opening; P_{atm} designating atmospheric pressure. For a color version of this figure, see www.iste.co.uk/sadchemloul/mechanics.zip*

The term π_2 is written as $\pi_2 = \frac{h}{d}$, but considering experience, we may write $\pi_2 = \frac{d}{h}$.

The dimension of the term π_3 gives us:

$$\begin{cases} M : \alpha + \gamma = 0 \\ L : -3\alpha + \beta - \gamma + 1 = 0 \\ T : -2\gamma - 2 = 0 \end{cases} \Rightarrow \begin{cases} \alpha = 1 \\ \beta = -1 \\ \gamma = -1 \end{cases}$$

The term π_3 is written as $\pi_3 = \frac{\rho g d}{P_o}$.

Considering the physical aspect of pressure, the physical quantity d is replaced with h.

Taking the experience into account, we may write:

$$\pi_3 = \frac{P_o}{\rho g d}$$

The dimension of the term π_4 gives us:

$$\begin{cases} M: \alpha + \beta + 1 = 0 \\ L: -3\alpha + \beta - \gamma - 1 = 0 \\ T: -2\gamma - 1 = 0 \end{cases} \Rightarrow \begin{cases} \alpha = -1/2 \\ \beta = -1 \\ \gamma = -1/2 \end{cases}$$

The term π_4 is written as $\pi_4 = \dfrac{\mu}{d\sqrt{\rho P_o}}$.

Taking the experience into account, we may write:

$$\pi_4 = \frac{d\sqrt{\rho P_o}}{\mu}$$

The terms in π are linked through:

$$F\left(\frac{Q_v}{d^2\sqrt{P_o/\rho}}, \frac{d}{h}, \frac{P_o}{\rho g d}, \frac{d\sqrt{\rho P_o}}{\mu}\right) = 0$$

The law that gives the flow is therefore written as:

$$Q_v = d^2\sqrt{P_o/\rho}\,F_1\left(\frac{d}{h}, \frac{P_o}{\rho g d}, \frac{d\sqrt{\rho P_o}}{\mu}\right)$$

Taking into account that h is very large in comparison with d, we may conclude that the relationship $\dfrac{d}{h}$ has no effect on the results, and therefore:

$$Q_v = d^2\sqrt{P_o/\rho}\,F_2\left(\frac{P_o}{\rho g d}, \frac{d\sqrt{\rho P_o}}{\mu}\right)$$

COMMENT 2.22.– *There is another solution to the problem. It can be considered that the flow Q_v is only a function of the pressure distribution that exists around the orifice, which is:*

$$P = P_o + \rho g h$$

Therefore, the physical quantities involved are Q_V, d, P_o, ρ and μ. The number of terms in π is reduced to 2 ($n-k=5-3=2$), which is:

$$\pi_1 = \frac{Q_V}{d^2\sqrt{P_o/\rho}} \text{ and } \pi_2 = \frac{d\sqrt{\rho P_o}}{\mu}$$

The expression of the flow rate in this case is:

$$Q_v = d^2\sqrt{\frac{P_o}{\rho}}f(d\frac{\sqrt{\rho P_o}}{\mu}) \text{ or } Q_v = d^2\sqrt{\frac{P}{\rho}+gh}f(d\frac{\sqrt{\rho(P+\rho gh)}}{\mu})$$

In this relationship, the terms used for pressure, P and ρgh, are grouped together so they are less general than the previous one used.

2.6.9. *Exercise 21: sudden narrowing of a section*

A fluid of density ρ and dynamic viscosity μ flows at a speed of V through a pipe of diameter D. This pipe narrows sharply to a cross-section of diameter d.

Determine the singular pressure drop relationship ΔP_s based on the problem data.

2.6.9.1. *Solution*

The relationship of the pressure drop is:

$$\Delta P_s = f(V, D, d, \rho, \mu)$$

The number of physical quantities is therefore $n = 6$.

The dimension of physical quantities is:

$$[\Delta P_s] = ML^{-1}T^{-2},\ [V] = LT^{-1},\ [D] = [d] = L,\ [\rho] = ML^{-3},\ [\mu] = ML^{-1}T^{-1}$$

The number of fundamental quantities is $k = 3$, and therefore there are three terms in π.

By taking V, D, and ρ as chosen or repeated quantities, the terms in π will have as their expression:

$$\begin{cases} \pi_1 = V^{\alpha} D^{\beta} \rho^{\gamma} \mu \\ \pi_2 = V^{\alpha} D^{\beta} \rho^{\gamma} d \\ \pi_3 = V^{\alpha} D^{\beta} \rho^{\gamma} \Delta P_s \end{cases}$$

The dimension of the term π_1 gives us:

$$\begin{cases} M: \gamma + 1 = 0 \\ L: \alpha + \beta - 3\gamma - 1 = 0 \\ T: -\alpha - 1 = 0 \end{cases} \rightarrow \begin{cases} \alpha = -1 \\ \beta = -1 \\ \gamma = -1 \end{cases}$$

$\pi_1 = \dfrac{\mu}{\rho V D}$ is the inverse of the Reynolds number, which can be written as:

$$\pi_1 = Re = \frac{\rho V D}{\mu}$$

In the same way, we obtain:

$$\pi_2 = \frac{d}{D} \text{ and } \pi_3 = \frac{\Delta P_s}{\rho V^2}$$

The terms in π verify the relationship $F(Re, \dfrac{d}{D}, \dfrac{\Delta P_s}{\rho V^2}) = 0$:

The relationship of the pressure drop is thus written as:

$$\Delta P_s = \rho V^2 F_1(Re, \frac{d}{D})$$

COMMENT 2.23.– *Experience shows that the pressure drop in a narrowing section is written as:*

$$\Delta P_S = k_S \frac{\rho V^2}{2}$$

where k_S *is the coefficient of the singular pressure drop* $k_S = F(Re, \frac{d}{D})$.

This last variable depends on the nature of the flow regime Re and on the shape and dimensions $\frac{d}{D}$ *of the singularity.*

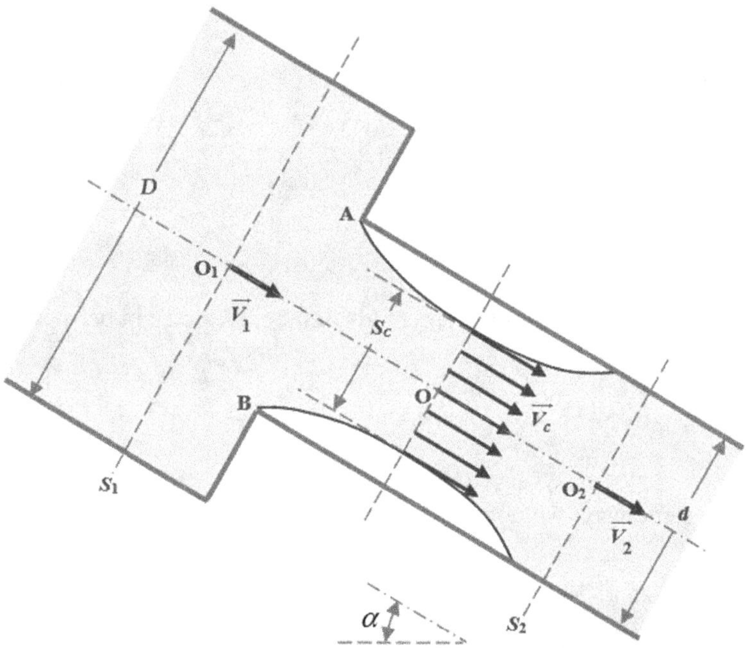

Figure 2.16. *Flow in a sharply narrowing section. For a color version of this figure, see www.iste.co.uk/sadchemloul/mechanics.zip*

Figure 2.16 provides a detailed depiction of the flow through a sharply narrowing section. Initially, the liquid vein will be detached at the points A and B, followed by:

– a contraction of the section of the liquid vein until reaching S_c *which is less than the section* S_2. *In this zone (the converging part), experience shows that the pressure drop is negligible, and the speed profile is uniform in the section* S_c*;*

– the liquid vein develops between the section S_c *and the section* S_2.

The general expression of the singular pressure drop created by the narrowing of a liquid vein is obtained using the theorem of the quantity of motion, and is given by Belanger's formula[18]*:*

$$\Delta P_s = \rho \frac{(V_c - V_2)^2}{2} = \rho \frac{V_2^2}{2}(\frac{1}{C_c} - 1)^2$$

with $C_c = \dfrac{S_c}{S_2}$ *being the coefficient of the contraction of the liquid vein's cross section.*

COMMENT 2.24.– *In the Belanger relationship, the output speed of the singularity is used. We can also use the input speed of the singularity.*

However, it should be noted that Belanger's formula is obtained using certain assumptions that do not fully reflect the true nature of the flow. This is why we sometimes use the modified formula devised by Barré de Saint Venant[19] *obtained from the experimental results:*

$$\Delta P_s = \rho \frac{(V_c - V_2)^2}{2} = \rho \frac{V_2^2}{2}\left[(\frac{1}{C_c} - 1)^2 + \frac{1}{9}\right]$$

Figure 2.17 shows the progression of the line of the load and the piezometric line through a sharply narrowing segment.

18 Jean Baptiste Charles Joseph Belanger, born on April 4, 1790, and who lived until May 8, 1874, was a mathematician who worked in the field of hydraulics and hydrodynamics.

19 Barré de Saint Venant, born on August 23, 1797, in the Château de Fortoiseau, in Villiers-en-Bière (Seine-et-Marne, France) and who lived until January 6, 1886, was an engineer, physicist, and mathematician.

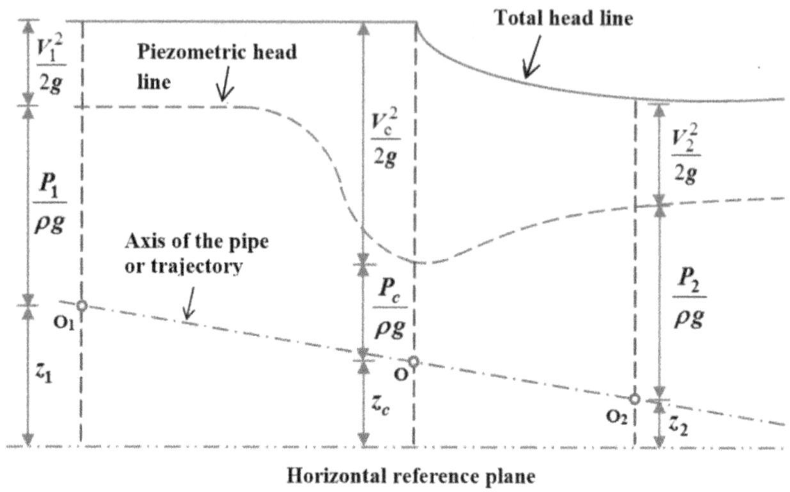

Figure 2.17. *Progression of the load line and the piezometric line through a sharply narrowing segment. For a color version of this figure, see www.iste.co.uk/sadchemloul/mechanics.zip*

At the level of the contracted section of the vein, the speed can reach its maximum value, while the pressure reaches its minimum value. It is therefore necessary for the absolute pressure in this section to be greater than the vapor tension of the liquid, so that the phenomenon of cavitation can be avoided.

2.6.10. *Exercise 22: capillary tube*

A capillary tube with radius r is submerged in a liquid of density ρ. Determine the relationship giving the height of the rise of the liquid in the capillary tube h and taking into account the surface tension σ of the liquid and the gravitational acceleration g.

2.6.10.1. *Solution*

The expression of the height of the rise is $h = f(r, \rho, \sigma, g)$.

The number of physical quantities is $n = 5$, and has as its dimensions:

$$[h] = [r] = L, [\rho] = ML^{-3}, [\sigma] = MT^{-2}, [g] = LT^{-2}$$

The number of fundamental quantities is $k=3$ and the number of terms in π is $n-k=2$.

If ρ, g and r are taken as chosen or repeated quantities, the terms in π are written as:

$$\begin{cases} \pi_1 = \rho^{\alpha} g^{\beta} r^{\gamma} h \\ \pi_2 = \rho^{\alpha} g^{\beta} r^{\gamma} \sigma \end{cases}$$

The dimension of the term π_1 gives us:

$$\begin{cases} M : \alpha = 0 \\ L : -3\alpha + \beta + \gamma + 1 = 0 \\ T : -2\beta = 0 \end{cases} \Rightarrow \begin{cases} \alpha = 0 \\ \beta = 0 \\ \gamma = -1 \end{cases}$$

and therefore $\pi_1 = \dfrac{h}{r}$.

The dimension of the term π_2 gives us:

$$\begin{cases} M : \alpha + 1 = 0 \\ L : -3\alpha + \beta + \gamma = 0 \\ T : -2\beta - 2 = 0 \end{cases} \Rightarrow \begin{cases} \alpha = -1 \\ \beta = -1 \\ \gamma = -2 \end{cases}$$

and therefore $\pi_2 = \dfrac{\sigma}{\rho g r^2}$.

The terms verify the relationship $F(\dfrac{h}{r}, \dfrac{\sigma}{\rho g r^2}) = 0$:

The expression of the height of the rise is:

$$\frac{h}{r} = F_1(\frac{\sigma}{\rho g r^2}) \text{ or } h = K\frac{\sigma}{\rho g r}$$

with K as a constant.

COMMENT 2.25.– *The height of rise is given by Jurin's law*[20]*:*

$$h = \frac{2\sigma}{\rho g r} = \frac{2\sigma cos\theta}{\rho g R}$$

with θ *being the angle of contact between the liquid and the tube wall, also called the juncture angle (see Figure 2.18).*

When immersing a glass capillary tube in a wetting liquid, such as water (see Figure 2.18), the liquid rises relative to the level of the open surface area of the container. The concave meniscus forms an angle $\theta < \frac{\pi}{2}$ *with the tube wall, and in the case of water* $\theta = 0$.

Because of this, it appears that a force exists that defies gravity: when we put the capillary tube in place, a curved surface forms, and in this case, Laplace's law[21] *dictates a difference in pressure between the liquid and the air. Thus, we see a decrease in pressure at the convex level of the meniscus. To compensate for this reduction in pressure, the liquid sees a rise in height h, caused by a hydrostatic pressure that is greater than atmospheric pressure (* $P > P_{atm}$ *).*

When the liquid does not dampen the walls of the capillary tube, that is, $\theta > \frac{\pi}{2}$ *, a capillary depression is observed (see Figure 2.19). This capillary depression is used in mercury porosimetry. This is a technique that allows the size distribution of pores to be determined. The principle of this is based on the fact that in order for mercury to penetrate the pores of a solid, it is necessary to apply a pressure so great that the pore size is reduced.*

20 James Jurin, born on December 15, 1684, and living until March 29, 1750, was an English doctor and physicist who left a legacy of pioneering works on capillary actions and epidemiology (inoculation).

21 Pierre-Simon de Laplace, born on March 23, 1749, in Beaumont-en-Auge (France) and who lived until March 5, 1827, was a mathematician, astronomer, physicist and political figure.

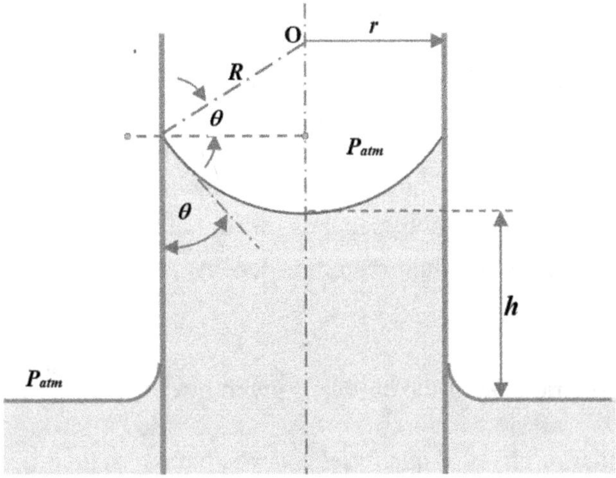

Figure 2.18. *Capillary rise in the case of water that dampens the solid from which the capillary tube is created. For a color version of this figure, see www.iste.co.uk/sadchemloul/mechanics.zip*

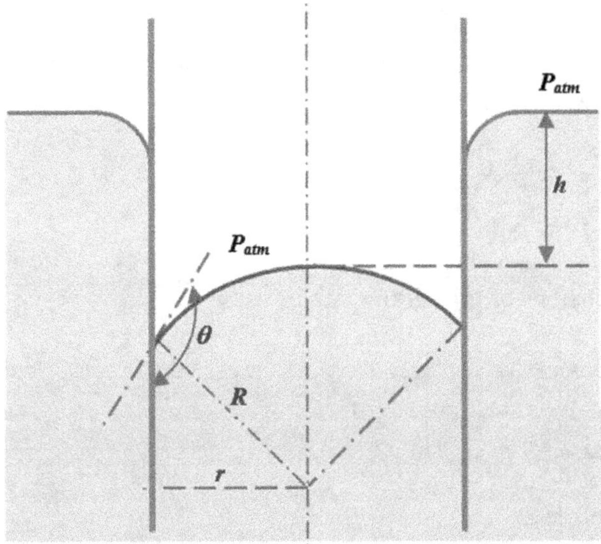

Figure 2.19. *Capillary depression in the case of mercury that does not dampen the solid from which the capillary tube is made. For a color version of this figure, see www.iste.co.uk/sadchemloul/mechanics.zip*

2.6.11. *Exercise 23: deformation of a bubble*

A bubble ascending in a fluid can remain spherical or be deformed by flattening its upper face. We may observe that the faster the bubbles move relative to the liquid, the less stable they become. Find a threshold value on a non-dimensional number above which the bubble becomes distorted. The deformation δ_b of the bubble depends on its radius *R*, its rate of ascent *V*, the surface tension of the fluid σ and its density ρ.

2.6.11.1. *Solution*

The deformation δ_b of the bubble is given by $\delta_b = f(R,\sigma,V,\rho)$.

The number of physical quantities is $n = 5$, having as their dimensions:

$$[\delta_b] = [R] = L,\ [V] = LT^{-1},\ [\sigma] = MT^{-2},\ [\rho] = ML^{-3}$$

The number of fundamental quantities is $n-k=3$ and therefore there are two terms in π.

Considering that σ, R, and ρ are chosen or repeated quantities, the terms π are written as:

$$\begin{cases} \pi_1 = \sigma^{\alpha} R^{\beta} \rho^{\gamma} \delta_b \\ \pi_1 = \sigma^{\alpha} R^{\beta} \rho^{\gamma} V \end{cases}$$

The dimensions of this term π_1 gives us:

$$\begin{cases} M : \alpha + \gamma = 0 \\ L : \beta - 3\gamma + 1 = 0 \\ T : -\ 2\alpha = 0 \end{cases} \Rightarrow \begin{cases} \alpha = 0 \\ \beta = -1 \\ \gamma = 0 \end{cases}$$

And therefore:

$$\pi_1 = \frac{\delta_b}{R}$$

This result is clear, because δ_b and R have the same dimension.

The dimensions of this term π_2 gives us:

$$\begin{cases} M: \alpha+\gamma=0 \\ L: \beta-3\gamma+1=0 \\ T: -2\alpha-1=0 \end{cases} \Rightarrow \begin{cases} \alpha=-1/2 \\ \beta=1/2 \\ \gamma=1/2 \end{cases}$$

And therefore:

$$\pi_2 = V\sqrt{\frac{\rho R}{\sigma}}$$

The terms verify the relationship:

$$F(\pi_1,\pi_2)=0 \text{ or } \delta_b = RF_1(\sqrt{\frac{\rho R V^2}{\sigma}})$$

COMMENT 2.26.– *The relative deformation of the bubble depends solely on the non-dimensional term* π_2. *The square of this term represents the Weber number. This number measures the ratio of the kinetic energy of the fluid to the energy of the bubble's distortion. In practice, the deformation of the bubble occurs beyond a critical value of the Weber number.*

Small Bond B_o *numbers are found in spherical bubbles. Their ability to be distorted increases with the Bond number.*

$$B_o = \frac{\rho g d^2}{\sigma} \quad or \quad B_o = \frac{\rho g d^3}{\sigma d} = \frac{\text{gravitational force}}{\text{surface tension force}}$$

For small capillary numbers, the forces of surface tension dominate. When the capillary number is zero, the deformation of a spherical bubble is zero. The variation of the shape of the bubbles on the basis of their volume depends on the level of viscosity. In this sense, the characterization of the different shapes of the bubbles is carried out by making a distinction between water and low viscosity Newtonian fluids as compared to high viscosity Newtonian fluids.

In viscous Newtonian fluids, bubbles pass through known shapes: spherical and then ellipsoid with vertically impacted revolutions, in which the back part of the bubble is flattened until it takes on a spherical, helmet-like shape for larger volumes.

For a Newtonian fluid, regardless of its size, the bubbles do not have a tail, but after reaching a certain volume, they are given a parachute-like shape (see Figure 2.20), as a depression forms at the back of the bubble.

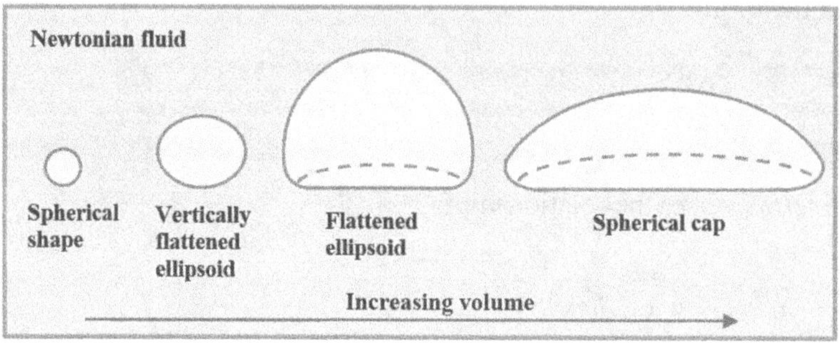

Figure 2.20. *Generic form of bubbles in a Newtonian fluid. For a color version of this figure, see www.iste.co.uk/sadchemloul/mechanics.zip*

2.6.12. *Exercise 24: laminar dynamic boundary layer on a flat plate*

Here, we will consider the laminar flow along a flat plate. The thickness of the dynamic boundary layer δ increases with the distance x at the bottom of the plate, and it also depends on the speed of the free flow U_∞, the dynamic viscosity of the fluid μ and the fluid's density ρ. Determine the expression of the dynamic boundary layer thickness δ.

2.6.12.1. *Solution*

The expression of the dynamic boundary layer thickness is:

$$\delta = f(x, U_\infty, \rho, \mu)$$

The number of physical quantities is $n = 5$, and has as its dimensions

$$[\delta] = [x] = L, [U_\infty] = LT^{-1}, [\rho] = ML^{-3}, [\mu] = ML^{-1}T^{-1}$$

The number of fundamental quantities is $k = 2$, and therefore there are two terms in π.

Considering that ρ, U_∞ and x are chosen or repeated quantities, the terms π are written as:

$$\begin{cases} \pi_1 = \rho^\alpha U_\infty^\beta x^\gamma \delta \\ \pi_2 = \rho^\alpha U_\infty^\beta x^\gamma \mu \end{cases}$$

The dimension of the term π_1 gives us:

$$\begin{cases} M: \alpha = 0 \\ L: -3\alpha + \beta + \gamma + 1 = 0 \\ T: -\beta = 0 \end{cases} \Rightarrow \begin{cases} \alpha = 0 \\ \beta = 0 \\ \gamma = -1 \end{cases}$$

and so $\pi_1 = \dfrac{\delta}{x}$, ratio between the thickness δ of the boundary layer and abscissa x where it is determined.

The dimension of the term π_2 gives us:

$$\begin{cases} M: \alpha + 1 = 0 \\ L: -3\alpha + \beta + \gamma + 1 = 0 \\ T: -\beta - 1 = 0 \end{cases} \Rightarrow \begin{cases} \alpha = -1 \\ \beta = -1 \\ \gamma = -1 \end{cases}$$

And thus $\pi_2 = \dfrac{\mu}{\rho U_\infty x}$ is the inverse of the Reynolds number according to the length of the flat plate, which is:

$$\pi_2 = Re_x = \frac{\rho U_\infty x}{\mu}$$

Dimensional analysis makes it possible to write the following:

$$F(\pi_1, \pi_2) = 0 \text{ or } \delta = xF_1(Re_x) = xF_1\left(\frac{\rho U_\infty x}{\mu}\right)$$

COMMENT 2.27.– *The simplest example of an external flow is that of a fluid under a pressure P_∞ which arrives at a uniform velocity U_∞ parallel to a flat wall (see Figure 2.21).*

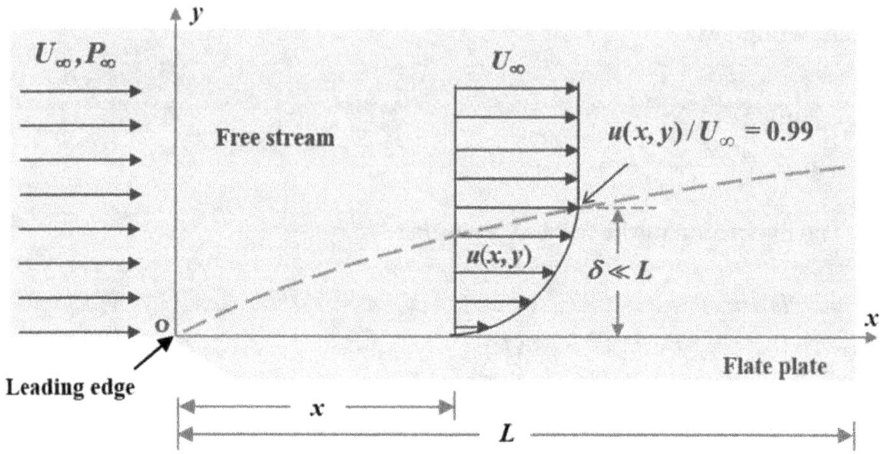

Figure 2.21. *Laminar dynamic boundary layer on a flat wall. For a color version of this figure, see www.iste.co.uk/sadchemloul/mechanics.zip*

When the fluid is in contact with the wall, the flow is divided into two regions:

– a region near the wall where a velocity gradient is established due to viscous forces: the particle-particle friction, and the particle-wall friction. In this region, the viscous forces are very large with respect to the inertia forces. The closer to the wall, the more the fluid is slowed, and the speed is cancelled out entirely along the surface of the wall (on the condition of adhesion to the wall). This region is known as the dynamic boundary layer;

– a region far from the wall where the velocity gradient is zero and the forces of inertia are greater than the viscous forces. This region is also referred to as healthy flow or Euler flow[22]*.*

The dynamic boundary layer is referred to as laminar when the Reynolds number is such that $Re_x < 10^5$.

22 Leonhard Euler, born on April 15, 1707, in Basel (Switzerland), and who lived until September 18, 1783, was a mathematician and physician.

The thickness of the boundary layer $\delta(x)$ *is conventionally defined by:*

$$\frac{u(x,y)}{U_\infty} = 0.99$$

NOTE 2.9.– *In order to determine the coefficient of friction due to the flow of the fluid on the wall, it is necessary to determine the velocity profile* $u(y)$. *This is given either by the exact solution, known as the "Blasius" solution, or by approximate solutions.*

The following table provides a comparison between the approximate solutions and the exact solution, with δ_d *being the displacement thickness of the boundary layer,* δ_{qm} *the thickness of the quantity of movement,* H_f *the shape factor that allows it to be determined whether or not the boundary layer will detach from the wall, the local coefficient of friction* C_f*, the overall coefficient of friction* C_f*, and* R_L *the Reynolds number calculated over the entire length of the flat wall.*

Profile $\frac{u}{U_\infty}$	$H_f = \frac{\delta_d}{\delta_{qm}}$	$\frac{\delta_d}{x}\sqrt{Re_x}$	$\frac{\delta_{qm}}{x}\sqrt{Re_x}$
Exact Solution	5	1.721	0.664
$\frac{y}{\delta}$	3.464	1.732	0.577
$2\frac{y}{\delta}-\left(\frac{y}{\delta}\right)^2$	5.48	1.83	0.730
$3\frac{y}{\delta}-\frac{1}{2}\left(\frac{y}{\delta}\right)^2$	4.64	1.74	0.646
$2\frac{y}{\delta}-2\left(\frac{y}{\delta}\right)^3+\left(\frac{y}{\delta}\right)^4$	5.84	1.752	0.686
$\sin\left(\frac{\pi}{2}\frac{y}{\delta}\right)$	1.741	1.741	0.654

Profile $\frac{u}{U_\infty}$	$\frac{\delta}{x}\sqrt{Re_x}$	$C_f\sqrt{Re_x}$	$C_F\sqrt{Re_L}$
Exact Solution	2.59	0.664	1.328
$\frac{y}{\delta}$	3.00	0.577	1.155
$2\frac{y}{\delta}-\left(\frac{y}{\delta}\right)^2$	2.50	0.730	1.460
$3\frac{y}{\delta}-\frac{1}{2}\left(\frac{y}{\delta}\right)^2$	2.70	0.646	1.292
$2\frac{y}{\delta}-2\left(\frac{y}{\delta}\right)^3+\left(\frac{y}{\delta}\right)^4$	2.55	0.686	1.372
$\sin\left(\frac{\pi}{2}\frac{y}{\delta}\right)$	2.66	0.654	1.310

Table 2.6. *Comparison between the exact solution and the approximate solutions in the case of a laminar dynamic boundary layer*

COMMENT 2.28.– *In the case of* $Re_x > 10^5$, *the dynamic boundary layer is turbulent (see Figure 2.22). The speed profile grows much more quickly with greater distance from the wall, and as a result, turbulent friction is much higher than laminar friction. Thus, the turbulent boundary layer is much more resistant to adverse pressure gradients, so it is more difficult to separate. In this turbulent boundary layer, two regions can be identified.*

1) The internal region. In this region, the velocity profile in the turbulent boundary layer is in a non-dimensional form $u^+ = f(y^+)$. *This region is formed by two zones (see Figure 2.23).*

Viscous sub-layer $Re_x < 5$ *or* 10 *: immediately next to the wall, the term laminar friction is dominant, as the average velocities and velocity fluctuations tend toward zero along the wall. The scale of speed that characterizes this zone is the speed of friction. The velocity profile is determined from the local mean velocity* $\bar{u}$, *the wall stress* τ_p, *the density* ρ, *and the dynamic viscosity μ of the fluid, and the perpendicular distance y to the wall.*

The expression of the local average speed is $\bar{u} = f(\tau_p, \rho, \mu, y)$.

The number of physical quantities is $n = 5$, and has as its dimensions:

$$\left[\bar{u}\right] = LT^{-1},\ \left[\tau_p\right] = ML^{-1}T^{-2},\ [\rho] = ML^{-3},\ [\mu] = ML^{-1}T^{-1},\ [y] = L$$

The number of fundamental quantities is $k = 3$. Thus, there are two terms in π.

Considering that τ_p, y and ρ are chosen or repeated quantities, the terms π are written as:

$$\begin{cases} \pi_1 = \tau_p^{\alpha} \rho^{\beta} y^{\gamma} \bar{u} \\ \pi_2 = \tau_p^{\alpha} \rho^{\beta} y^{\gamma} \mu \end{cases}$$

The dimension of the terms in π gives the following expressions:

$$\pi_1 = \frac{\bar{u}}{\sqrt{\tau_p / \rho}} = \frac{\bar{u}}{u_f} \text{ and } \pi_2 = \frac{\mu}{y\rho\sqrt{\tau_p / \rho}} = \frac{\mu}{y\rho u_f}$$

The term $u_f = \sqrt{\tau_p / \rho}$ is called the "friction velocity" because it has the same dimension as a velocity and is determined from the stress along the wall (friction). The term π_1 is generally written as $\pi_1 = \frac{\bar{u}}{u_f} = u^+$. The term π_2 has no dimension and represents the inverse of the Reynolds number, but a dimensional analysis allows us to write:

$$\pi_2 = \frac{\rho u_f y}{\mu} = y^+$$

The two terms in π confirm the relationship:

$$F(u^+, y^+) = 0 \; or \; u^+ = F_1(y^+)$$

The form of the function F_1 depends on the flow behavior.

For $y^+ < 1$, we have $u^+ = y^+$.

– Buffer zone $5 < y^+ < 30$: the connection of the linear law that is valid in the viscous sublayer and the logarithmic profile that is valid in the upper part of the inner layer occurs gradually in a buffer region.

– Logarithmic zone (barrier law) $30 < y^+ < 500$: the friction between the threads of the flow primarily occur because of turbulent agitation, and the average local velocity is given by:

$$u^+ = 5.6 log_{10} y^+ + 5$$

2) The external region. In this region, there are two zones:

– Velocity deficit zone $y^+ > 500$: as we leave the viscous sub-layer, the term of turbulent friction gradually becomes predominant with respect to laminar friction, such that as we approach the edge of the boundary layer, the laminar friction becomes negligible. The difference between the average velocity $\bar{u}$ in this area and the outer velocity U_∞ is on the order of magnitude of the friction velocity u_f:

$$\frac{U_\infty - \bar{u}}{u_f} = 5.6 log_{10} \frac{y}{\delta} - 2.5$$

– The wake zone, where movements are very disjointed. This is an important cause of the dissipation of energy. It defines the profile of velocities through a completely empirical law:

$$\frac{\bar{u}}{u_f} = 13.65 log\left(\frac{y u_f}{\nu}\right) + 2.48\left(2\sin^2(\frac{\pi y}{\delta})\right) + 4.9$$

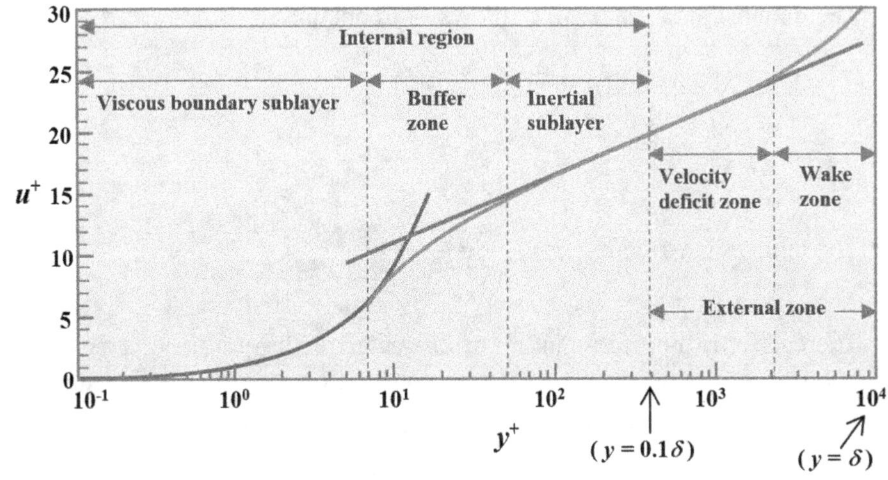

Figure 2.22. *Velocity profile through the turbulent boundary layer of a flat plate. For a color version of this figure, see www.iste.co.uk/sadchemloul/mechanics.zip*

2.6.13. *Exercise 25: power of a stirrer*

Determine the power p transmitted by a stirrer to a liquid of density ρ. The stirrer of diameter D spins at the rotational speed N.

2.6.13.1. *Solution*

The power transmitted by the stirrer to the liquid is given by the relationship $p = f(\rho, N, D)$.

The number of physical quantities is $n = 5$, and has as its dimensions:

$$[p] = ML^{-2}T^{-3},\ [D] = L,\ [\rho] = ML^{-3},\ [N] = T^{-1}$$

The number of fundamental quantities is $n = 3$, and therefore there is one term in π. Considering that D, ρ and N are chosen or repeated quantities, the terms π are written as $\pi = D^{\alpha}\rho^{\beta}N^{\gamma}p$.

The dimension of the term π allows us to obtain:

$$\begin{cases} M: \beta + 1 = 0 \\ L: \alpha - 2\beta + 2 = 0 \\ T: -\gamma - 3 = 0 \end{cases} \begin{cases} \alpha = -5 \\ \beta = -1 \\ \gamma = -3 \end{cases}$$

The term π is written as $\pi = \dfrac{p}{\rho N^3 D^5}$.

There is only one term in π to characterize the problem, it is not a function of any other, it is necessarily constant ($\pi = constant = K$), and the relationship of the power is written as $p = K\rho N^3 D^5$.

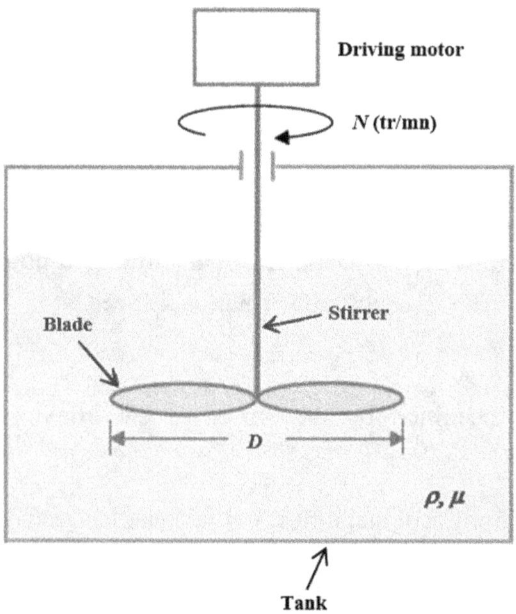

Figure 2.23. *Scheme of a stirrer for liquids. For a color version of this figure, see www.iste.co.uk/sadchemloul/mechanics.zip*

COMMENT 2.29.– *The power* p *transferred by a stirrer to a liquid depends on the flow scheme:*

– in a turbulent regime ($Re = \frac{\rho N D^2}{\mu} < 10^4$ *), the power consumed by the stirring within this turbulent regime is* $p = (C - C_o)\omega$, *with* μ *being the dynamic viscosity of the liquid,* C *the torque transferred to the liquid,* C_o *the torque while not subjected to a load (in the air) and* N *the rotating speed of the stirrer;*

– for the turbulent regime the power number $N_p = \frac{p}{\rho N^3 D^5} = K$ *characterizing the energy consumption is a constant for a given agitator tank. The constant* K *therefore represents the power number;*

– in the laminar scheme $(Re < 10)$ *the product* $N_p.Re$ *is constant.*

Stirring is achieved by putting the liquid phase into movement. If the mass is made up of several components, the stirring tends to homogenize the mass in contact with the reagents, so that they react more efficiently. Most of the stirring and mixing operations are carried out by means of a tilting stirrer that rotates around a shaft placed in a vessel, which is generally cylindrical in shape. If no stirring is done, the mixture follows Fick's law[23] *(molecular diffusion).*

The role of the stirrer is to cause greater movement of the fluid, and therefore the agitation due to forced convection. To accomplish this, different types of stirrers have been devised, each of which has a design adapted to a given application.

There are two main classes of agitation propellers, following the movement of fluids generated in the tank relative to the axis of rotation of the mobile (see Figure 2.24):

– axial flow mobiles, which create a flow of fluid in the tank from the top to the bottom;

– radial flow mobiles that create radial flow perpendicular to the rotation axis.

---

23 Adolf Eugene Fick, born on September 3, 1829, in Kassel (Germany), and living until August 21, 1901, was a physiologist. In 1851, he became a doctor of medicine. In 1855, he introduced two phenomenological laws addressing the diffusion of matter, known as Fick's laws.

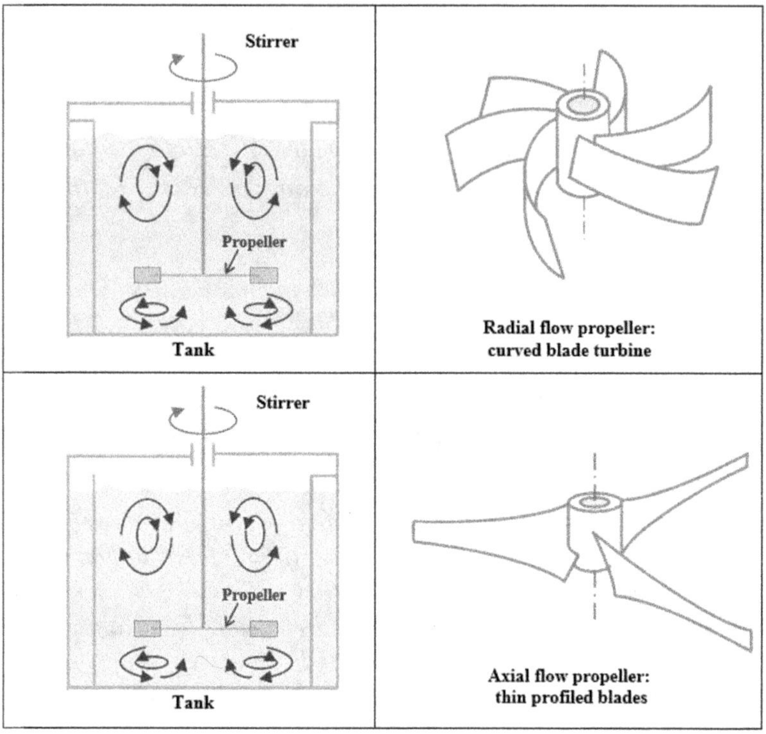

Figure 2.24. *Example of axial and radial flow propellers. For a color version of this figure, see www.iste.co.uk/sadchemloul/mechanics.zip*

3

Similarity of Flows

3.1. Definition and principle of similarity

Similarity allows us to determine whether two flows are comparable, and if they are, to determine the relationships that make it possible to transpose the numerical results from one to the other.

3.1.1. *Geometric similarity*

If L_1 and L_2 are the reference lengths representing homologous dimensions of the two domains, then their relationship:

$$\lambda_L = \frac{L_2}{L_1} \quad [3.1]$$

is the geometric scale ratio (or length scale ratio) between the prototype (1) and the model (2).

For two homologous points M_1 and M_2, the ratios relative to each of the coordinates are equal to λ_L:

$$\lambda_L = \frac{x_2}{x_1} = \frac{y_2}{y_1} = \frac{z_2}{z_1} \quad [3.2]$$

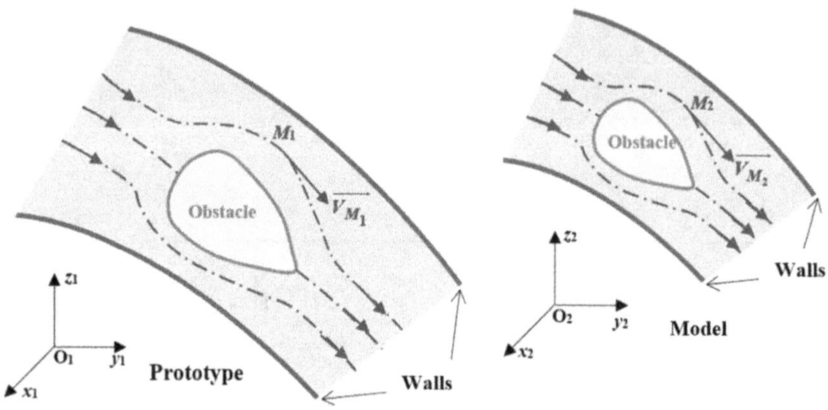

Figure 3.1. *Similarity of two flows. For a color version of this figure, see www.iste.co.uk/sadchemloul/mechanics.zip*

NOTE 3.1.– *It is impossible to obtain a complete geometric similarity, because it is not possible to create a model that perfectly represents the state of the surface for the walls of a structure.*

3.1.2. *Kinematic similarity*

This similarity concerns the streamlines and the trajectories, and consequently, the velocity field.

Two flows are kinematically similar when, regardless of the homologous point pairs (M_1 and M_2)[1] considered, the relationship $\vec{V_2}/\vec{V_1}$ of the velocities at these points remains constant. The directions of these two velocities are therefore the same with respect to the reference trihedrons, and the ratio of their module is independent of the two points considered.

If λ_V and λ_t represent the scale ratio of velocity and times, respectively:

$$\lambda_V = \frac{\vec{V_2}}{\vec{V_1}}, \ \lambda_t = \frac{t_2}{t_1} \qquad [3.3]$$

1 Index 1, where p is assigned to the prototype, and index 2, where m is assigned to the model.

The relationship between the distance, velocity, and time allows us to obtain:

$$\lambda_L = \lambda_V \lambda_t \tag{3.4}$$

COMMENT 3.1.– *This basic relationship shows that the three scale ratios of lengths, velocity, and times cannot be chosen independently.*

For the two flows to be fully comparable, the above property of the velocity field must be extended to the other flow characteristics (field of pressure and temperature) and to the unique properties of the fluid (density, viscosity coefficient).

If it is possible to define the corresponding scale ratios, which could mean, for example, the ratio $\lambda_\rho = \dfrac{\rho_2}{\rho_1}$ *is independent of the two points (*M_1 *and* M_2*), the two flows are said to be in "complete similarity".*

The final element of comparison involves the conditions of the boundaries, involving the geometry or particular values of velocities, pressures, or temperatures. These values, taken from the prototype and the model, must confirm the valid scale ratios within the flow.

3.1.3. *Dynamic similarity*

To establish the conditions of similarity in fluid mechanics, only the forces of inertia, gravity, and friction are taken into account. We will now consider the flow of an incompressible and viscous fluid subject to a gravitational field. It is governed by the Navier-Stokes equations, which are written in vector form:

$$\frac{\partial \vec{V}}{\partial t} + (\vec{V}.\overrightarrow{grad})\vec{V} = \vec{F_v} - \frac{1}{\rho}\overrightarrow{grad}P + \nu\Delta\vec{V} \tag{3.5}$$

We will make these equations non-dimensional by introducing the new variables:

$$x' = \frac{x}{L_R},\ y' = \frac{y}{L_R}, z' = \frac{z}{L_R},\ \vec{V'} = \frac{\vec{V}}{V_R},\ t' = \frac{t}{t_R},\ P' = \frac{P}{\rho V_R^2} \tag{3.6}$$

L_R, V_R and t_R, are reference quantities (length, velocity and time) selected from the flow under consideration.

Generally, in fluid mechanics, the volume force which intervenes is that of gravity, and therefore $\overrightarrow{F_v} = g\vec{k}$. Thus, we obtain:

$$\frac{L_R}{t_R V_R}\left(\frac{\partial \overrightarrow{V'}}{\partial t'}\right) + (\overrightarrow{V'}.\overrightarrow{grad'})\overrightarrow{V'} = \frac{L_R}{V_R^2} g\vec{k} - \overrightarrow{grad'}P' + \frac{\nu}{L_R V_R}(\Delta'\overrightarrow{V'}) \qquad [3.7]$$

NOTE 3.2.– *The primes (′) assigned to the operators $\overrightarrow{grad'}$ and Δ' indicate the derivatives taken with respect to the dimensional coordinates x', y' and z'.*

The ratio $\frac{L_R}{t_R V_R}$ is the Strouhal number[2] $S = \frac{L_R}{t_R V_R}$, the ratio $\frac{\nu}{L_R V_R}$ is the inverse of the Reynolds number $Re = \frac{V_R L_R}{\nu}$, and the ratio $\frac{L_R}{V_R^2}$ is the inverse of the square of the Froude number $Fr^2 = \frac{V_R^2}{L_R}$.

The relationship [3.7] is thus written as:

$$S\left(\frac{\partial \overrightarrow{V'}}{\partial t'}\right) + (\overrightarrow{V'}.\overrightarrow{grad'})\overrightarrow{V'} = \frac{L_R}{V_R^2} g\vec{k} - \overrightarrow{grad'}P' + \frac{1}{Re}(\Delta'\overrightarrow{V'}) \qquad [3.8]$$

Now, we will apply this equation to the two flows, prototype (1) and model (2) of Figure 3.1:

$$S_1\left(\frac{\partial \overrightarrow{V_1'}}{\partial t_1'}\right) + (\overrightarrow{V_1'}.\overrightarrow{grad_1'})\overrightarrow{V_1'} = \frac{L_{R_1}}{V_{R_1}} g_1\vec{k} - \overrightarrow{grad_1'}P_1' + \frac{1}{Re_1}(\Delta_1'\overrightarrow{V_1'}) \qquad [3.9]$$

$$S_2\left(\frac{\partial \overrightarrow{V_2'}}{\partial t_2'}\right) + \left(\overrightarrow{V_2'}.\overrightarrow{grad_2'}\right)\overrightarrow{V_2'} = \frac{L_{R_2}}{V_R^2} g_2\vec{k} - \overrightarrow{grad_2'}P_2' + \frac{1}{Re_2}(\Delta_2'\overrightarrow{V_2'}) \qquad [3.10]$$

2 Vincent Strouhal, born on April 10, 1850, in Seĉ (Czech Republic) and who lived until January 26, 1922, was a physicist.

We will choose the reference quantities so that they are homologous quantities of the two flows. Their respective ratios are equal to the corresponding scale ratio, that is:

$$\frac{x_2}{x_1} = \frac{x_2' L_{R_2}}{x_1' L_{R_2}} = \lambda_L$$

or $\lambda_L = \dfrac{L_{R_2}}{L_{R_1}}$ and thus $x_1' = x_2'$.

The same reasoning shows that all variables are then in equal pairs:

$$x_1' = x_2',\ y_1' = y_2',\ z_1' = z_2',\ \overrightarrow{V_1'} = \overrightarrow{V_2'},\ t_1' = t_2',\ P_1' = P_2'$$

Thus, for equations [3.9] and [3.10] to have the same solutions, it is necessary for the coefficients of the different terms of these two equations to be in equal pairs. These equal values constitute the conditions of similarity between the two flows, which are:

$$S_1 = S_2,\ \frac{L_{R_1}}{V_{R_1}^2} = \frac{L_{R_2}}{V_{R_2}^2},\ Re_1 = Re_2 \qquad [3.11]$$

Table 3.1 gives the three conditions of similarity.

	Similarity condition of Strouhal	Similarity condition of Reynolds	Similarity condition of Froude
Dimensionless numbers	$S_1 = S_2$	$Re_1 = Re_2$	$Fr_1 = Fr_2$
Reference properties ratio	$\dfrac{L_{R_1}}{V_{R_1} t_{R_1}} = \dfrac{L_{R_2}}{V_{R_2} t_{R_2}}$	$\dfrac{V_{R_1} L_{R_1}}{\nu_1} = \dfrac{V_{R_2} L_{R_2}}{\nu_2}$	$\dfrac{V_{R_1}}{\sqrt{g_1 L_{R_1}}} = \dfrac{V_{R_2}}{\sqrt{g_2 L_{R_2}}}$
Scale relationships	$\lambda_L = \lambda_V \lambda_t$	$\lambda_L \lambda_V = \lambda_\nu$	$\lambda_V = \lambda_L^{1/2} \lambda_g^{1/2}$

Table 3.1. *Conditions of similarity between two flows*

3.1.4. *Similarity conditions for viscous, incompressible, non-heavy fluids (Reynolds similarity)*

The condition of similarity between two viscous, incompressible and non-heavy fluid flows is the equality of the Reynolds numbers. These numbers must be calculated with homologous quantities.

If the ratios λ and λ_V are known, the different scales relating to the similarity of two viscous, non-compressible and lightweight fluids are written as:

Scale of	Different fluids	Same fluid
Velocities	$\lambda_L = \lambda_V^{-1} \lambda_v$	$\lambda_V = \lambda_L^{-1}$
Times	$\lambda_t = \lambda_L^2 \lambda_V^{-1}$	$\lambda_t = \lambda_L^2$
Pressures	$\lambda_{P^*} = \lambda_L^{-2} \lambda_\rho \lambda_V^2$	$\lambda_{P^*} = \lambda_L^{-2}$ [3]
Flow rates	$\lambda_Q = \lambda_L \lambda_V$	$\lambda_Q = \lambda_L$
Hydraulic powers	$\lambda P_h = \lambda_L^{-1} \lambda_\rho \lambda_V^3$	$\lambda P_h = \lambda_L^{-1}$

Table 3.2. *Reynolds similarity conditions*

3.1.5. *Similarity conditions for non-viscous, incompressible, heavy fluids (Reech-Froude similarity)*

These are free surface flows where the inertial forces are relatively large and the dimensions of the structure (prototype) and model large enough so that the viscous frictional forces existing along the boundary layers of the wall are negligible. The similarity condition of Reech-Froude, which represents the equality of the Froude numbers, is given by Table 3.3. These Froude numbers must be calculated with homologous quantities. By introducing scale ratios, this condition is written:

$$\lambda_L \lambda_g = \lambda_V^2 \qquad [3.12]$$

Generally, with λ_g being always equal or very close to 1 (same position), there are three unknowns: λ_L, λ_V, and λ_t.

3 The piezoelectric pressure P^*, often used in hydraulics, is the sum of the pressure at a point P and the pressure from the altitude $\rho g z$ and the pressure at a point $P^* = P + \rho g z$.

If the length scale λ_L is chosen, a similarity is possible, and only in one form. The different scales of similarity are given in Table 3.3.

Scale of	Different positions and different fluids $\lambda_g \neq 1$ and $\lambda_\rho \neq 1$	Same position and same fluid $\lambda_g = 1$ and $\lambda_\rho = 1$
Velocities	$\lambda_V = \lambda_L^{1/2} \lambda_g^{1/2}$	$\lambda_V = \lambda_L^{1/2}$
Times	$\lambda_t = \lambda_L^{1/2} \lambda_g^{-1/2}$	$\lambda_t = \lambda_L^{1/2}$
Pressures	$\lambda_{P*} = \lambda_L \lambda_g \lambda_\rho$	$\lambda_{P*} = \lambda_L$
Flow rates	$\lambda_Q = \lambda_L^{5/2} \lambda_g^{1/2}$	$\lambda_Q = \lambda_L^{1/2}$
Hydraulic powers	$\lambda_{P_h} = \lambda_L^{7/2} \lambda_g^{3/2} \lambda_\rho$	$\lambda_{P_h} = \lambda_L^{7/2}$

Table 3.3. *Reech-Froude conditions of similarity*

3.1.6. *Similarity requirements for non-viscous, non-compressible, heavy fluids*

If a fluid is considered both viscous and heavy, the equality of the Reynolds and Froude numbers should be checked simultaneously:

$$Re_1 = Re_2, \ Fr_1 = Fr_2 \tag{3.13}$$

The three conditions of similarity, those of Strouhal, Reynolds, and Reech-Froude, are given by the relationships of the following system, respectively:

$$\begin{cases} \lambda_L = \lambda_V \lambda_t \\ \lambda_L \lambda_V = \lambda_\nu \\ \lambda_L \lambda_g = \lambda_V^2 \end{cases} \tag{3.14}$$

By eliminating the velocity scale ratio λ_V between the last two relationships, we obtain:

$$\lambda_L = \lambda_\nu^{2/3} \lambda_g^{-1/3} \tag{3.15}$$

NOTE 3.3.– *Similarity will therefore be possible only if the geometric scale ratio* λ_L *confirms this relationship: we must adapt* λ_L *to the fluid used. The relationship* [3.15] *is only feasible in certain very special cases, which demonstrates that in practice, the Reynolds and Froude conditions are incompatible.*

3.1.7. Conditions of similarity of turbulent flows

In a turbulent flow, new forces appear due to the disorderly agitation of the particles. These forces are frictional forces. They are expressed from Reynolds stresses, which are the time-averaged values of the product of two of the components of the velocity of fluctuation, like the term $\overline{u'v'}$[4] (Reynolds tensor). Experience shows that these constraints can be expressed as a function of the components of the average particle velocity.

The condition is met if the flows of the structure (prototype) and the model have corresponding Reynolds numbers that are both greater than a certain threshold value of the flow under consideration.

If the fluids are not heavy, there is no condition of similarity in the truest sense. We are left with $\lambda_L = \lambda_V \lambda_t$, and we can arbitrarily choose both the geometric scale ratio and one of the ratios relating to the flow, such as that of the piezometric pressures $P*$. Thus, we have:

Scale of	Different fluids	Same fluid
Velocities	$\lambda_V = \lambda_{p*}^{1/2}\lambda_\rho^{-1/2}$	$\lambda_V = \lambda_{p*}^{1/2}$
Times	$\lambda_t = \lambda_L \lambda_{p*}^{-1/2}\lambda_\rho^{-1/2}$	$\lambda = \lambda_L \lambda_{p*}^{-1/2}$
Flow rates	$\lambda_Q = \lambda_L^2 \lambda_{p*}^{1/2}\lambda_\rho^{-1/2}$	$\lambda_Q = \lambda_L^2 \lambda_{p*}^{1/2}$
Hydraulic powers	$\lambda_{P_h} = \lambda_L^2 \lambda_{p*}^{3/2}\lambda_\rho^{-1/2}$	$\lambda_{P_h} = \lambda_L^2 \lambda_{p*}^{3/2}$

Table 3.4. *Conditions of similarity of turbulent flows*

4 u' and v' are the following fluctuating speeds, x and y, respectively.

3.1.8. *Distortion of the model*

When the dimensions of the structure are very different, depending on the direction considered, it is no longer possible to apply the same geometric scale ratio to all the dimensions of the structure. In this case, two different geometric scaling ratios must be selected: one λ_L that applies to planar dimensions, and the other λ_H that applies to the vertical dimensions. We then perform a distorted similarity. The ratio λ_H / λ_L is called the "distortion of the model".

The velocity varies as the square root of the change in piezometric pressure $P*$ or as the square root of the change in height of the free surface. The comparison of velocities is therefore made using the vertical scale ratio, that is, $\lambda_V = \lambda_H^{1/2}$. Taking into account the relationship $\lambda_L = \lambda_V \lambda_t$, we can deduce the following expressions for the other scale ratios (with the same fluid in the structure and the model):

Scale of	Same fluid
Velocities	$\lambda_V = \lambda_H^{1/2}$
Times	$\lambda_t = \lambda_L \lambda_H^{-1/2}$
Flow rates	$\lambda_Q = \lambda_L \lambda_H^{3/2}$

Table 3.5. *Conditions of similarity in the case of the distortion of the model*

3.2. Exercises: similarity of flows

3.2.1. *Exercise 1: similarity between ships*

Two geometrically similar ships, whose geometric scale ratio is 1:15, move through the same body of water with density ρ.

1) What would the relationship between the velocities be?

2) Calculate the ratio of the resistance to advancement of the ship (neglecting the effect of the Reynolds number).

3) Calculate the ratio of the powers necessary for the advancement of the ship.

4) What is the ratio of Reynolds numbers?

5) The movement resistance of the large vessel is calculated from the test results on the reduced model. Will the real resistance be higher or lower than that which is calculated theoretically? Why?

3.2.1.1. *Solutions*

1) The flow of water around the ship is a free surface flow, so the Reech-Froude similarity condition is used. With L_n being one of the dimensions of the ship and V being its speed:

$$Fr_m = Fr_p \text{ or } \frac{V_m}{\sqrt{g_m L_{n,m}}} = \frac{V_p}{\sqrt{g_p L_{n,p}}}$$

with:

$$\lambda_L = \frac{L_{n,m}}{L_{n,p}}, \lambda_V = \frac{V_m}{V_p}, \lambda_g = \frac{g_m}{g_p}$$

This last scale, which takes into account the acceleration of gravity, is equal to the unit, since it involves the same position or location. The condition of similarity is written as:

$$\lambda_V = \lambda_L^{1/2} = \left(\frac{1}{15}\right)^{1/2} = 0.25$$

2) Where Fr is the Force of resistance to the movement of the ship due to the friction of the water against the ship, given by $Fr = C_f \rho \frac{V^2}{2} S$, with C_f being the coefficient of friction, and S being the friction of the water against the ship.

With the two ships considered to be in similarity, we may write:

$$\frac{Fr_m}{Fr_p} = \frac{C_{f,m}}{C_{f,p}} \frac{\rho_m}{\rho_p} \frac{V_m^2}{V_p^2} \frac{S_m}{S_p}$$

NOTE 3.4.– $\frac{C_{f,m}}{C_{f,p}} = 1$ *because the coefficient of friction depends on the Froude number, and since this number is the same for the model and the prototype, the friction coefficients of the model and the prototype are equal.*

$\frac{\rho_m}{\rho_p} = 1$, *it is the same fluid (water).*

$$\frac{S_m}{S_p} = \left(\frac{L_m}{L_p}\right)^2 = \frac{1}{\lambda_L^2}, \frac{V_m}{V_p} = \frac{1}{\lambda_V^2}$$

Taking into account question one, the relationship between the frictional forces F_f is therefore written as:

$$\frac{F_{f,m}}{F_{f,p}} = \frac{1}{\lambda_L^2}\frac{1}{\lambda_V^2} = \frac{1}{\lambda_L^3} = \frac{1}{15^3} = \frac{1}{3,375}$$

3) The power p required for the ship to advance is given by the product of the movement resistance force or friction F_f and the speed of advancement V, or $p = F_f V$.

The ratio of powers is:

$$\frac{p_m}{p_p} = \frac{F_{f,m}}{F_{f,p}}\frac{V_m}{V_p} = \frac{1}{\lambda^3}\frac{1}{\lambda^{1/2}} = \frac{1}{15^{7/2}} = \frac{1}{13,071.31}$$

4) the ratio of the Reynolds numbers is:

$$\frac{Re_m}{Re_p} = \frac{V_m}{V_p}\frac{L_{n,m}}{L_{n,p}}\frac{\nu_p}{\nu_m} = \lambda_V \lambda_L \lambda_\nu^{-1} = \lambda_V^{3/2}\lambda_\nu^{-1}$$

The scale of kinematic viscosities is $\lambda_v = 1$, because the same fluid (water) is used. The Reynolds number ratio has the value of:

$$\frac{Re_m}{Re_p} = \lambda_V{}^{3/2} = \left(\frac{1}{15}\right)^{3/2} = \frac{1}{58.09}$$

5) The actual resistance is lower than that which is calculated theoretically, since viscosity forces are more significant for the model.

NOTE 3.5.– *The movement of a ship is influenced by two parameters: the wake, and in particular, the drag that the water exerts on the hull. The latter of these two can be broken down into six elements: shape drag, friction drag, wave impact drag, wave creation drag, drift-induced drag and steering drag. The most important of these are the shape drag, characterized by the drag coefficient which depends on the Reynolds number* $C_t = f(Re)$, *and the wave drag, characterized by a wave coefficient which depends on the Froude number* $C_{vague} = f(Fr)$.

To create a model in similarity, it would be best to choose a scale of reduction for all the geometric dimensions and to have the same Reynolds number and the same Froude number to fully recreate the relative shares of shape drag and wave drag. We can easily verify that this is impossible, at the very least, if water is still used as a supporting fluid and the gravitational acceleration is not changed.

When traction tests are carried out on a model in a hull testing basin, the best results may be obtained by working either with Reynolds similarity or with Froude similarity.

3.2.2. *Exercise 2: similarity of centrifugal pumps*

A centrifugal pump delivers $1{,}000\ liters/mn$ by working against a manometric height of 15 m when the rotational speed is $1{,}500\ rpm$. The diameter of the rotor is 30 cm and the power at the brake is $6\ CV$[5]. A geometrically similar pump with an impeller diameter of 35 cm must operate

5 One horsepower (*HP*) is the power required to vertically raise a weight of 75 *kg* to 1 *m* in height in 1 second = 75 *kgm*/s = 735.4990000873 *W*. Horsepower is kilogram-force per cm^2.

at $1,750\ rpm$. Assuming the efficiencies equal, determine for this second pump:

1) the height supplied to the water;

2) the flow of water pumped;

3) the hydraulic power supplied to the water.

3.2.2.1. *Solutions*

REVIEW.– In a centrifugal pump, the mechanical energy of the impeller is converted into hydraulic energy. The functioning of a centrifugal pump is based on Euler's principle. The hydraulic energy the water receives is equal to the difference between the output energy (discharge) and the input energy (suction) of the pump. It is also occasionally called the "net discharge head" (net energy received by the water right at the outlet of the pump).

When the dimensions of the pump are small, and the suction and discharge section are equal, it can be shown that the hydraulic energy received by the water at the pump's output is equal to the difference in pressure. This difference in pressure can be measured simply by using two manometers: one placed at the inlet of the pump, and the other at the outlet.

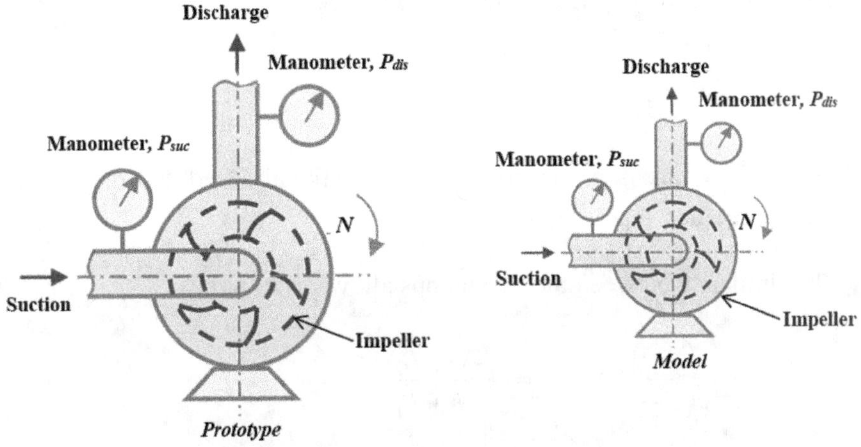

Figure 3.2. *Model and prototype of a centrifugal pump. For a color version of this figure, see www.iste.co.uk/sadchemloul/mechanics.zip*

The data of the characteristics of two pumps are given in the table below:

Characteristics	Pump 1	Pump 2
Flow rate	$Q_1 = liters\,/\,mn$	$Q_2 = ?$
Manometric height	$H_1 = 15\ m$	$H_2 = ?$
Rotation speed	$N_1 = 1{,}500\ rpm$	$N_2 = 1{,}750\ rpm$

Table 3.6. *Data of the characteristics of pumps in similarity*

In order for two pumps to be in similarity, they must have the same overall efficiency.

1) The manometric height H is proportional to the rotational speed and diameter of the pump impeller, that is, $H \sim N^2 D^2$.

With the two pumps considered to be in similarity, we may write:

$$\frac{H_1}{H_2} = \left(\frac{N_1}{N_2}\right)^2 \left(\frac{D_1}{D_2}\right)^2 \quad \text{or} \quad H_2 = H_1 \left(\frac{N_2}{N_1}\right)^2 \left(\frac{D_2}{D_1}\right)^2$$

The height load supplied to the second pump:

$$H_2 = 15 \times \left(\frac{1,750}{1,500}\right)^2 \times \left(\frac{35}{30}\right)^2 = 27.78\ m$$

2) The flow rate is proportional to the rotational speed and diameter of the pump impeller, that is, $Q \sim ND^3$.

The similarity between the two pumps allows us to write:

$$\frac{Q_1}{Q_2} = \frac{N_1}{N_2}\left(\frac{D_1}{D_2}\right)^3 \quad \text{or} \quad Q_2 = Q_1 \frac{N_2}{N_1}\left(\frac{D_2}{D_1}\right)^3$$

The water flow rate output by the second pump is:

$$Q_2 = 1,000 \times \left(\frac{1,750}{1,500}\right)^3 \times \left(\frac{35}{30}\right)^3 = 1,852.62\ liters\,/\,s$$

3) The hydraulic power supplied by the impeller of the pump to the liquid is given by the relationship $p_h = \rho g Q H$. It therefore depends on the nature of the liquid, and the diameter and the speed of rotation of the pump impeller, that is, $p_h \sim \rho g N^3 D^5$.

The similarity between the two pumps allows us to write:

$$\frac{p_{h1}}{p_{h2}} = \frac{\rho_1}{\rho_2}\frac{g_1}{g_2}\frac{N_1^3}{N_2^3}\frac{D_1^5}{D_2^5}$$

Since this involves the same fluid $\frac{\rho_1}{\rho_2} = 1$ and the same position $\frac{g_2}{g_1} = 1$, the power supplied by the second pump is:

$$p_{h2} = p_{h1}\left(\frac{N_2}{N_1}\right)^3\left(\frac{D_2}{D_1}\right)^5 = 6\times\left(\frac{1,750}{1,500}\right)^3\times\left(\frac{35}{30}\right)^5 = 20.59\ CV$$

COMMENT 3.2.– *This exercise can also be solved by using the manometric coefficient* $\delta_H = \frac{gH}{N^2D^2}$, *the flow rate coefficient* $\mu_Q = \frac{Q}{ND^3}$ *and the power coefficient* $\tau_p = \frac{p}{\rho N^3 D^5}$.

These coefficients, called the "Rateau coefficients", are dimensionless numbers that characterize a hydraulic machine. They are created from the geometric and mechanical characteristics of the machine. For a given pump, the characteristic described using the Rateau coefficients is always the same, regardless of the rotational speed and fluid. Another factor to consider is the efficiency coefficient, also called overall efficiency of the centrifugal pump:

$$\eta = \frac{\text{hydraulic power delivred to the fluid at the output pump}}{\text{mechanical power absorbed by the pump impeller shaft (power input pump)}}$$

The Rateau coefficients predict the behavior of a pump (characteristics or working parameters) on the basis of the characteristics of a pump from the same family and that operates at a different speed. The Rateau coefficients are useful for determining similarities between pumps. Essentially, from a given characteristic, we will be able to determine the functionality of the pump for other impeller speeds, other fluids, or other impeller dimensions.

The functioning of two machines is governed solely by the flow rate coefficient $\mu_Q = \dfrac{Q}{ND^3}$, *and the other groups are equal to each other. In this case, it is easy to see how the different sizes vary. The operational parameters of identical machines and different machines are given in Table 3.7.*

	Machines, geometrically similar	**Machines are different, geometrically similar, and carrying the same fluid**	**Identical machines carrying the same fluid**	
			Rotating at different rotational speeds	**Rotating at the same speed and with different impeller blades**
The overall efficiency of the pump η	does not vary	does not vary	does not vary	does not vary
The flow rate Q **is proportional to**	N and R^3	N and R^3	N	R
The manometric height H **is proportional to**	R^2 and N^2	R^2 and N^2	N^2	R^2
The hydraulic power P_h **is proportional to**	ρ, N^3 and R^5	N^3 and R^5	N^3	R^5
The torque C **is proportional to**	ρ, N^2 and R^5	N^2 and R^5	N^2	R^5

Table 3.7. *Working parameters of the same machines and different machines*

The table below provides some power levels using horsepower.

Unit	Symbol	Definition
Watt	W	1 J/s, named after the engineer James Watt[6]
Kilogram meter per second	kg.m/s	1 kg.m/s or 9.80665 W
Huygens	H_y	1 kg.m/s or 9.80665 W named after the scholar Christian Huygens[7]
CUNA horsepower	CUNA HP[8]	735.49875 W, these initials stand for *commissione unificazione normalizzazione autoveicoli*, formerly used by Italian automakers
DIN horsepower	DIN HP[9]	735.79875 W these initials stand for *deutsche industrie normen*, formerly used by German automakers
Fiscal horsepower	cv or CV	The Fiscal or administrative horsepower is not a unit of power. It determines the fiscal bracket of a motor vehicle and is expressed in fiscal horsepower. As of July 1, 1998: $p_{CV} = [rejet\ CO_2(g/km)/45] + (p_{kW}/40)^{1,6}$
SAE horsepower	SAE HP[10]	735.49875 W the initials stand for *society of automotive engineers*, formerly used by American automakers
Steam horsepower	HP	735.49875 W, the power required to lift approximately 75 kg a height of 1m for 1s
var	var	1 W

Table 3.8. *Different uses of power*

6 James Watt, born on January 19 , 1736, in Greenock (Scotland) and who died on August 25, 1819, was an engineer whose improvements to the steam engine were considered one of the key breakthroughs of the industrial revolution.

7 Christian Huygens, born on April 14, 1629 in The Hague (Netherlands) and who died on July 8, 1695, was a mathematician, astronomer, and physicist.

8 CUNA horsepower: the power of an engine on a testing rig without any accessories. BHP: the British use their own unit, equal to 1.0139 HP.

9 DIN horsepower: the power transmitted to the drive shaft of a vehicle in running order.

10 SAE horsepower: the power of an engine on a test rig with a minimum number of accessories (no alternator, fan, etc.), greater than the DIN power. SAE could be interpreted as: “without any equipment”.

3.2.3. *Exercise 3: volumetric pumps with small dimensions*

To predict the behavior of a small oil pump, tests are conducted on a model using air. The oil pump must be driven by a motor with power $0.05\ CV$ rotating at a speed of 1,800 *rpm*. We are given a motor with power $0.25\ CV$ to rotate the air pump at a rotational speed of 600 *rpm*. Given that the density of the oil is 0.912 and the density of air (constant) is $\rho = 1.225\ kg/m^3$, what is the value of the geometric scale ratio $\frac{D_{model}}{D_{prototype}}$?

3.2.3.1. *Solution*

The data of the model and prototype pump data are shown in the table below:

Characteristics	Model pump	Prototype pump
Rotation speed	$N_m = 600\ rpm$	$N_p = 1{,}800\ rpm$
Manometric height	$p_m = 0.25\ CV$	$p_p = 0.05\ CV$
Density	$\rho = 1.225\ kg/m^3$	$\rho_p = 912\ kg/m^3$
Geometric scale ratio	$\frac{D_{model}}{D_{prototype}} = ?$	

Table 3.9. *Data of the characteristics of pumps in similarity*

The manometric power of the model pump and the prototype pump is:

$$p_p = \rho g Q_p H_p \text{ and } p_m = \rho g Q_m H_m$$

Since it involves the same location $\frac{g_m}{g_p} = 1$, the scale ratio of the powers is therefore written as:

$$\lambda_{ph} = \lambda_\rho \lambda_Q \lambda_H$$

By neglecting the hydraulic head losses between the impeller and the pump outlet, the the manometric height H becomes equal to the theoretical height H_{th} given by the Euler relation:

$$H = H_{th} = \frac{U_2 C_{2u} - U_1 C_{1u}}{g}$$

with:

– U_1 and U_2 the peripheral speeds at the inlet and outlet of the pump impeller, respectively;

– C_{1u} and C_{2u} the tangential components of the absolute velocity of the fluid at the input and output of the pump impeller, respectively;

– the manometric height H is proportional to the diameter D and rotation speed N of the wheel:

$$H \sim UC_u \sim U^2 \sim N^2 D^2$$

The scale ratio of the manometric height is proportional to the scale of the rotational speeds and the geometric scale:

$$\lambda_H = \frac{H_m}{H_p} = \left(\frac{N_m}{N_p}\right)^2 \left(\frac{D_m}{D_p}\right)^2 = \lambda_N^2 \lambda_L^2$$

NOTE 3.6.– *Not considering the flow of leakage between the impeller and the wall of the front of the volute, the internal flow in the impeller is equal to the useful flow at the outlet of the pump. This last value is determined by the product of the radial component of the absolute velocity* C_r *C_r and the flow passage section* S_p. *The radial component of the absolute velocity is proportional to the diameter D and speed of rotation N of the impeller.*

With B being the height of the blade of the pump impeller, we obtain:

$$\begin{cases} C_r \sim ND \\ S_p \sim B\pi D \sim D^2 \end{cases} \Rightarrow Q \sim ND^3$$

The flow scale ratio is:

$$\lambda_Q = \frac{Q_m}{Q_p} = \frac{N_m}{N_p}\left(\frac{D_m}{D_p}\right)^3 = \lambda_N \lambda_L^3$$

The scale ratio of powers is written as:

$$\lambda_p = \frac{p_m}{p_p} = \frac{\rho_m}{\rho_p}\frac{Q_m}{Q_p}\frac{H_m}{H_p} = \lambda_\rho \lambda_N^3 \lambda_L^5$$

The scale of geometric similarity is $\lambda_L = \left(\frac{\lambda_p}{\lambda_\rho \lambda_N^3}\right)^{1/5}$ and has the value of:

$$\lambda_L = \left(\frac{0.25}{0.025} \times \frac{912}{1.225} \times \left(\frac{1,800}{600}\right)\right)^{1/5} = 10.01$$

3.2.4. *Exercise 4: characteristics of a centrifugal pump*

A centrifugal pump generates a flow rate of $Q_v = 0.070\ m^3 / s$ under a manometric height of $H = 7.5\ m$ and a rotational speed $N = 1,450\ rpm$, and requires a power of $p = 9CV$. If the speed is reduced to $N = 1,200\ rpm$, calculate the flow rate, manometric height, and power assuming that the efficiencies are equal.

3.2.4.1. *Solution*

The characteristics of the centrifugal pump for both rotational speeds are given in the table below:

Since the pump has two speeds of rotation, the law of similarity is applied.

The flow scale ratio gives us:

$$\frac{Q_1}{Q_2} = \frac{N_1}{N_2}\left(\frac{D_1}{D_2}\right)^2$$

Rotation speed	$N_1 = 1{,}450\ rpm$	$N_2 = 1{,}200\ rpm$
Flow rate	$Q_1 = 0.070\ m^3/s$	$Q_2 = ?$
Manometric height	$H_1 = 7\ m$	$H_2 = ?$
Power	$p_1 = 9\ CV$	$p_2 = ?$
Overall efficiency	The overall efficiency of the pump is constant for both rotational speeds	

Table 3.10. *Data of the characteristics of pumps in similarity*

Since it involves the same pump, and therefore the same impeller $\frac{D_2}{D_1} = 1$, the scale ratio is written as:

$$\frac{Q_2}{Q_1} = \frac{N_2}{N_1} = \lambda_N$$

That is:

$$\lambda_N = \frac{1{,}200}{1{,}450} = 0.827$$

The flow rate of the model centrifugal pump at the rotational speed $N_1 = 1450\ rpm$ is:

$$Q_2 = 0.827 Q_1 = 0.827 \times 0.070 = 0.0579\ m^3/s$$

The scale ratio of the manometric heights is:

$$\lambda_H = \frac{H_2}{H_1} = \left(\frac{N_2}{N_1}\right)^2 \underbrace{\left(\frac{D_2}{D_1}\right)^2}_{=1} = \lambda_N^2$$

The manometric height at the rotation speed $N_2 = 1{,}200\ rpm$ is:

$$H_2 = \lambda_N^2 H_1 = 0.827^2 \times 7.5 = 5.13\ m$$

The scale ratio of the hydraulic powers at the pump outlet is:

$$\lambda_p = \frac{p_2}{p_1} = \frac{\rho_2}{\rho_1}\frac{g_2}{g_1}\frac{Q_2}{Q_1}\frac{H_2}{H_1}$$

Since the fluids $\frac{\rho_2}{\rho_1} = 1$ are the same and are in the same place $\frac{g_2}{g_1} = 1$, the scale ratio of the powers is written:

$$\lambda_p = \frac{Q_2}{Q_1}\frac{H_2}{H_1} = \lambda_N \lambda_N^2 = \lambda_N^3$$

The hydraulic power at the rotation speed $N_2 = rpm$ is:

$$p_2 = \lambda_N^3 p_1 = 0.827^3 \times 9 = 5.09\ CV$$

COMMENT 3.3.– *In this case, it is the same pump coupled to a motor, such as an electric motor, with several rotational speeds. It is therefore possible to determine experimentally the characteristics of the pump for a single rotational speed. For the other rotational speeds, the characteristics of the pump are determined by applying the similarity of flows in centrifugal pumps. We can directly apply Rateau's laws of similarity.*

3.2.5. *Exercise 5: test of an automobile in a wind tunnel*

An automobile must be tested in a wind tunnel. The speed of the car on which the test is conducted is 108 km/h. The dimensions of the wind tunnel do not allow for a full-size car to be tested, so we create a geometrically similar model of a height of 1 m instead of 1.5 m for the prototype. What is the air velocity in *m/s* at which the test should be carried out?

3.2.5.1. *Solution*

The data of the model and of the prototype are shown in the table below:

We apply the Reynolds similarity condition:

$$Re_m = Re_p \text{ or } \frac{V_m h_m}{\nu_m} = \frac{V_p h_p}{\nu_p}$$

Characteristics of the automobile	Model	Prototype
Speed	$V_m = ?$	$V_p = 108\ km/h$
Height	$h_m = 1\,m$	$h_p = 1.5\,m$

Table 3.11. *Data of the characteristics of automobiles in similarity*

By considering the scale ratios, we obtain $\lambda_V \lambda_L = \lambda_\nu$.

Since the same fluid $\lambda_\nu = 1$ is used, then $\lambda_\nu = \lambda_L^{-1}$.

Thus, the speed of the model is:

$$V_m = V_p \lambda_V = V_p \lambda_L^{-1}$$

That is:

$$V_m = 1.5 \times 102 = 153\ km/h = 42.5\ m/s$$

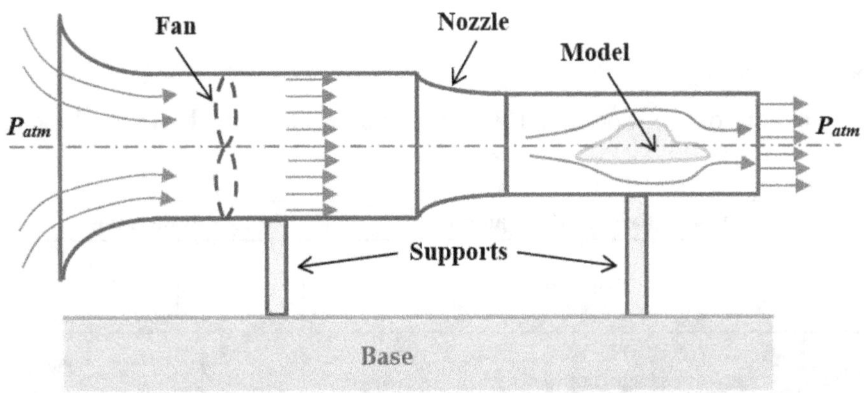

Figure 3.3. *Diagram of a wind tunnel. For a color version of this figure, see www.iste.co.uk/sadchemloul/mechanics.zip*

COMMENT 3.4.– *A wind tunnel is a test facility used in aerodynamics to study a flow of air around a body (car, airplane), usually a model of a smaller size than the full version.*

In a wind tunnel, forces can be measured and flows can be seen that are often impossible to achieve under real-world conditions of movement.

In a wind tunnel, we can fairly accurately determine the behavior of a car in motion, or the flight of an aircraft that is a model of a full-size version. It is much easier, cheaper, and safer to build and test a mock-up than to build and operate a car or fly a real airplane.

A wind tunnel makes it possible to study the aerodynamic phenomena induced by the flow of air around a vehicle. The knowledge of these phenomena allows us to reduce the consumption made by vehicles by decreasing their drag, to improve their behavior on the road by influencing their lift force (or their downforce) and to reduce aeroacoustic phenomena and turbulence at high speeds.

3.2.6. Exercise 6: power ratios (p_{model} / $p_{prototype}$) of a pump

What is the hydraulic power ratio of a prototype centrifugal pump and its scale model $\frac{1}{5}$, if the ratio of manometric loads or heights $\lambda_H = \frac{H_m}{H_p}$ is $\frac{1}{4}$?

3.2.6.1. *Solution*

The data of the model and prototype pump data are shown in the table below:

Characteristics of the pump	Scale ratios
Diameters	$\lambda_L = \frac{D_m}{D_p} = \frac{1}{5}$
Heights or loads	$\lambda_H = \frac{H_m}{H_p} = \frac{1}{4}$
Hydraulic powers	$\frac{p_m}{p_p} = ?$

Table 3.12. *Ratios of scales between the model pump and the prototype pump*

Since the fluids $\frac{\rho_m}{\rho_p} = 1$ are the same and are in the same place $\frac{g_m}{g_p} = 1$, the scale ratio of the hydraulic powers is written:

$$\lambda_p = \frac{\rho_m}{\rho_p}\frac{g_m}{g_p}\frac{Q_m}{Q_p}\frac{H_m}{H_p} = \lambda_\rho \lambda_g \lambda_Q \lambda_H = \lambda_Q \lambda_H$$

Not considering the hydraulic pressure losses between the impeller and the pump outlet, the manometric height H is equal to the theoretical height H_{th} given by the Euler relationship:

$$H = H_{th} = \frac{U_2 C_{2u} - U_1 C_{1u}}{g}$$

with:

– U_1 and U_2 the peripheral speeds at the inlet and outlet of the pump impeller, respectively;

– C_{1u} and C_{2u} the tangential components of the absolute velocity of the fluid at the input and output of the pump impeller, respectively;

– the manometric height H is proportional to the diameter D and rotation speed N of the impeller:

$$H \sim UC_u \sim U^2 \sim N^2 D^2$$

The scale ratio of the manometric heights is proportional to the scale of the rotational speeds and at the geometric scale:

$$\lambda_H = \left(\frac{N_m}{N_p}\right)^2 \left(\frac{D_m}{D_p}\right)^2 = \lambda_N^2 \lambda_L^2$$

The flow is determined by the product of the radial component C_r of the absolute velocity and the flow passage section S_p. The radial component of the absolute speed is proportional to the diameter and rotation speed of the impeller.

$$\begin{cases} C_r \sim ND \\ S_p \sim B\pi D \sim D^2 \end{cases} \Rightarrow Q \sim ND^3$$

with B being the height of the blades.

The flow scale ratio is:

$$\lambda_Q = \frac{Q_m}{Q_p} = \frac{N_m}{N_p}\left(\frac{D_m}{D_p}\right)^3 = \lambda_N \lambda_L^3$$

The scale ratio of powers is written as:

$$\lambda_p = \lambda_\rho \lambda_g \lambda_Q \lambda_H$$

Since this involves the same fluid ($\lambda_\rho = 1$) and the same position ($\lambda_g = 1$), we obtain:

$$\lambda_p = \lambda_Q \lambda_H$$

Taking the scale of rotational speeds λ_N from the scale of manometric height λ_H and placing it onto the scale of powers λ_p, we obtain:

$$\lambda_p = \lambda_Q \lambda_H = \lambda_N^3 \lambda_L^5 = (\lambda_L^{-1} \lambda_H^{1/2})^3 \lambda_L^5 = \lambda_L^2 \lambda_H^{3/2}$$

$$\lambda_p = \left(\frac{1}{5}\right)^2 \times \left(\frac{1}{4}\right)^{3/2} = 200$$

The power of the model is 200 times the power of the prototype.

3.2.7. *Exercise 7: flow in a pipe*

A flow of water at 20 °C in a pipe with a diameter of 5 *cm* has an average velocity of 0.2 *m/s*.

1) Calculate the Reynolds number of the flow. What is the flow regime?

2) A flow of oil with dynamic viscosity $\mu = 0.05\ PI$ and density $\rho_e = 900\ kg/m^3$, similar to the previous one, is carried out in a pipe with a diameter of 30 *cm*. Calculate the volumetric flow rate of oil.

3.2.7.1. *Solutions*

1) At a temperature of 20 °C, the tables of the thermo-physical properties of water give us $\nu = 1.006 \times 10^{-6}\ m^2/s$.

The Reynolds number is:

$$Re = \frac{VD}{\nu} = \frac{0.2 \times 0.05}{1.006 \times 10^{-6}} = 9{,}940$$

Thus, the regime is turbulent.

We may recall that the Reynolds number is dimensionless and represents the ratio of the forces of inertia to the forces of gravity.

2) The characteristics of the flows of water and oil are given in Table 3.13:

Characteristics	Fluid	
	Water	Oil
Velocity of flow	$V_e = 0.2 m/s$	$V_h = ?$ and $Q_h = ?$
Diameter of the pipe	$D_e = 5\ cm$	$D_h = 30\ cm$
Kinematic viscosity	$\nu_e = 1.006 \times 10^{-6}\ m^2/s$	$\nu_h = \frac{\mu_h}{\rho_h} = \frac{0.05}{900} = 5.55 \times 10^{-5} m^2/s$

Table 3.13. *Data on the characteristics of water and oil flows in similarity*

The flow of water and oil are in similarity if the Reynolds similarity condition is confirmed, that is:

$$Re_e = Re_h$$

or $\dfrac{V_e D_e}{\nu_e} = \dfrac{V_h D_h}{\nu_h}$:

The flow speed of the oil is:

$$V_h = V_e \frac{\nu_h}{\nu_e} \frac{D_e}{D_h} = 0.2 \times \frac{5.55 \times 10^{-5}}{1.006 \times 10^{-6}} \times \frac{5}{30} = 1.84\ m/s$$

The flow rate is:

$$Q_h = V_h \frac{\pi D_h^2}{4} = 1.84 \times \frac{\pi (0.3)^2}{4} = 0.130\ m^3/s$$

3.2.8. *Exercise 8: viscous force on a rotating disk*

We would like to study the torque C required to rotate a disc of diameter D at an angular speed ω in a fluid of density ρ and dynamic viscosity μ.

1) Use dimensional analysis to establish an expression of the torque C (assuming a perfectly smooth disk).

2) A disc 22.5 *cm* in diameter rotating at 1,400 *rpm* in water at $20°C$ requires a torque force of 1.1 *Nm*. We would like to provide for the operation of a similar disk, 67.5 *cm* in diameter, in the air, with a dynamic viscosity $\mu = 17 \times 10^{-6}\ PI$ and density $\rho = 1.125\ kg/m^3$.

a) What is the speed at which this disk should rotated to achieve similarity with the previous case?

b) Then, what would be the power p required to propel it?

3.2.8.1. *Solutions*

1) to establish the expression for the torque $C = f(\rho, \omega, D, \mu)$, we will apply the Vaschy-Buckingham theorem. The dimensions of the physical quantities involved in this physical phenomenon are:

$$[C] = ML^{-2}T^{-2},\ [\rho] = ML^{-3},\ [\omega] = T^{-1}, [D] = L, [\mu] = ML^{-1}T^{-1}$$

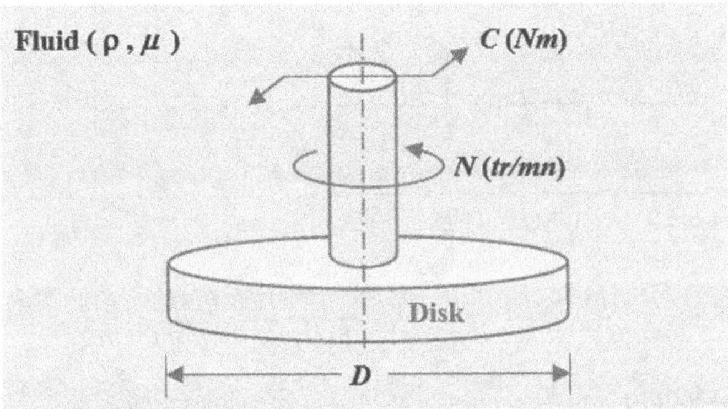

Figure 3.4. *Disk rotating in a viscous fluid. For a color version of this figure, see www.iste.co.uk/sadchemloul/mechanics.zip*

The number of physical quantities is $n = 5$, that of fundamental quantities or basic dimensions is $k = 3$, so there are two terms in π.

To form the terms in π by taking ρ, ω, and D as repeating variables:

$$\begin{cases} \pi_1 = \rho^{\alpha} \omega^{\beta} D^{\gamma} C \\ \pi_2 = \rho^{\alpha} \omega^{\beta} D^{\gamma} \mu \end{cases}$$

The dimension of the term in π_1:

$$\begin{cases} M: \alpha + 1 = 0 \\ L: -3\alpha + \beta + 1 = 0 \\ T: -\beta - 2 = 0 \end{cases} \Rightarrow \begin{cases} \alpha = -1 \\ \beta = -2 \\ \gamma = 5 \end{cases}$$

and therefore $\pi_1 = \dfrac{C}{\rho \omega^2 D^5}$

The same applies to the term π_2:

$$\begin{cases} M:\alpha+1=0 \\ L:-3\alpha+\beta+1=0 \\ T:-\beta-1=0 \end{cases} \Rightarrow \begin{cases} \alpha=-1 \\ \beta=-1 \\ \gamma=-2 \end{cases}$$

and therefore $\pi_2 = \dfrac{\mu}{\rho\omega D^2}$.

NOTE 3.7.– *The term π_2 can be written as:*

$$\pi_2 = \frac{\mu}{\rho\omega D^2} = \frac{\mu}{2(\rho\omega D/2)D} = \frac{\mu}{2VD}$$

It is therefore proportional to the inverse of the Reynolds number. Dimensional analysis makes it possible to write the following:

$$\pi_2 = f(Re)$$

The two terms in π verify the relationship:

$$F(\pi_1, \pi_2) = 0$$

Hence the expression for the torque required to drive the disk in the fluid is:

$$C = \rho\omega^2 D^5 F_1(Re)$$

Other expressions of the couple are possible if the repeating variables are changed.

2a) The flow characteristics of the prototype and model are described in the table below.

Characteristics of the disk	Prototype: water	Model: air
Diameter	$D_p = 22.5\ cm$	$D_m = 67.5\ cm$
Rotation speed	$N_p = 1{,}400\ rpm$	$N_m = ?$
Angular velocity	$\omega_p = \frac{2\pi N_p}{60} = \frac{2\times\pi\times 1{,}400}{60} = 146.53\ rd/s$	$\omega_m = ?$
Propulsion torque	$C_p = 1.1\ Nm$	$C_m = ?$
Propulsion power	$p_p = \omega_p C_p = 146.53\times 1.1 = 161.18\ W$	$p_m = ?$

Table 3.14. *Data on the characteristics of the viscous force exerted by a fluid (water and air) on a rotating disk*

We will apply the condition of Reynolds similarity[11] between the rotation of the disk in the water and in the air:

$$Re_m = Re_p$$

either $\frac{V_m D_m}{\nu_m} = \frac{V_p D_p}{\nu_p}$ or $\frac{\omega_m D_m^2}{\nu_m} = \frac{\omega_p D_p^2}{\nu_p}$.

At a temperature of 20°C, the tables of the thermophysical properties of water (prototype) give us $\nu_p = 1.006\times 10^{-6}\ m^2/s$. The angular speed of the disk in the air, or of the air particle bonded to the disk, is:

$$\omega_m = \omega_p \frac{\nu_m}{\nu_p}\frac{D_p^2}{D_m^2} = \omega_p \frac{\mu_m / \rho_m}{\nu_p}\frac{D_p^2}{D_m^2}$$

$$\omega_m = 146.53\times\frac{(17\times 10^{-6}/1.125)}{1.006\times 10^{-6}}\times\left(\frac{22.5}{67.5}\right)^2 = 24.55\ rd/s$$

11 The Reynolds number that is considered is that of the fluid particle (water or air), the velocity U in contact with the disk.

The speed of the rotation of the disk in the air is:

$$N_m = \frac{60\omega_m}{2\pi} = \frac{60 \times 244.55}{2 \times \pi} = 2{,}336.54\ rpm$$

2b) With the two flows being in similarity, the flows of the model and prototype have the same Reynolds number, and thus the value of the function $F_1(Re)$ remains constant:

$$Re_m = Re_p,\ F_1(Re) = F_2(Re) \text{ or } \frac{C_m}{\rho_m \omega_m^2 D_m^5} = \frac{C_p}{\rho_p \omega_p^2 D_p^2}$$

The friction torque applied by the air to the disk is:

$$C_m = C_p \frac{\rho_m}{\rho_p}\left(\frac{\omega_m}{\omega_p}\right)^2\left(\frac{D_m}{D_p}\right)^5$$

$$C_m = 1.1 \times \frac{1.125}{10^3} \times \left(\frac{244.55}{146.53}\right)^2 \times \left(\frac{67.5}{22.5}\right)^5 = 0.837\,Nm$$

The power required to propel the disk in the air is therefore:

$$p_m = C_m \omega_m = 0.837 \times 244.55 = 204.68\,Nm$$

COMMENT 3.5.– *The example of practical use of such a system is a disperser, which consists in the incorporation of ultrafine solid particles in liquid mediums, causing colloidal suspensions (such as in paints). These are characterized by the fact that solid particles dispersed at a high speed can no longer settle, due to the gravitational field of the earth.*

The best possible dispersion will depend on the following:

– the relationship between the dimensions of the pot and the diameter between both the peripheral speeds and the depth of immersion of the disk, which must be chosen correctly;

– the rheological behavior of the materials to be crushed.

3.2.9. *Exercise 9: development study of a hydroelectric gallery*

Before opening up the flow of water, we would like know the pressure drops to be expected in the gallery of a hydroelectric development which has just been built. For this purpose, we plan to measure losses occurring in a flow of air blown through the same gallery.

1) Since the expected flow rate of water is $7.5\ m^3/s$, what air flow rate should be blown through to meet the conditions of similarity?

2) On a measurement section of 500 *m*, equal losses were obtained with this air flow, $12\ mbars$. What will be the value of the head losses of the gallery with a length of $6{,}050\ m$?

The kinematic viscosities and volumetric masses of water and air are given:

– air: $\rho_{air} = 1.25\ kg/m^3$; $\nu_{air} = 14.70 \times 10^{-6}\ m^2/s$

– water: $\rho_{water} = 10^3\ kg/m^3$; $\nu_{water} = 1.15 \times 10^{-6}\ m^2/s$

3.2.9.1. *Solutions*

1) The necessary air flow rate through the gallery, in order for the flows of water and air to be in similarity, shall be determined by applying Reynolds similarity.

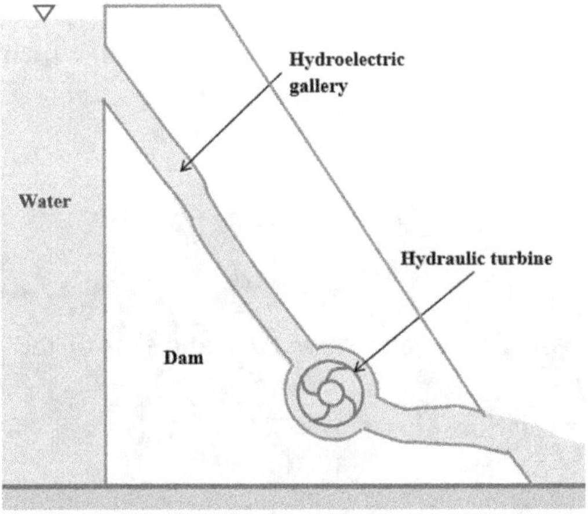

Figure 3.5. *Hydroelectric development gallery. For a color version of this figure, see www.iste.co.uk/sadchemloul/mechanics.zip*

The data of the prototype and of the model are shown in the table below:

Characteristics	Prototype: water	Model: air
Kinematic viscosity of the fluid	$\nu_p = 1.15 \times 10^{-6}\ m^2/s$	$\nu_m = 14.70 \times 10^{-6}\ m^2/s$
Density of the fluid	$\rho_p = 10^3\ kg/m^3$	$\rho_m = 1.25\ kg/m^3$
Flow rate of fluid	$Q_p = 7.5\ m^3/s$	$Q_m = ?$
Length of the gallery	$L_p = 6{,}050\ m$	$L_m = 500\ m$
Pressure losses in the gallery	$\Delta P_p = ?$	$\Delta P_m = 12\ mbar$

Table 3.15. *Data on the characteristics of the flow through in the hydroelectric development gallery*

We will apply the condition of Reynolds similarity:

$$Re_m = Re_p \text{ or } \frac{V_m D_m}{\nu_{mp}} = \frac{V_p D_p}{\nu_p}$$

With $V = \dfrac{4Q}{\pi D^2}$ we obtain $\dfrac{Q_m}{\nu_m D_m} = \dfrac{Q_p}{\nu_p D_p}$.

Since we are working with the same passage (same diameter of $\dfrac{D_m}{D_p} = 1$), the air flow is:

$$Q_m = Q_p \frac{\nu_m}{\nu_p} = 7.5 \times \frac{14.70 \times 10^{-6}}{1.15 \times 10^{-6}} = 95.86\ m^3/s$$

2) The head loss of the water flow ΔP_p and that of the air flow ΔP_m are determined by the relationship of $\Delta P = \lambda \dfrac{\rho V^2 L}{2D}$.

With $\lambda = f(Re, \frac{\varepsilon}{D})$, the coefficient of the linear pressure drop depends on the flow regime (Reynolds number Re) and the condition of the inner surface of the pipe (relative roughness $\frac{\varepsilon}{D}$).

NOTE 3.8.– *The Reynolds number and relative roughness are the same for the two flows in the same hydroelectric gallery. It is therefore necessary to apply the Reynolds similarity condition, which results in a constant value of the coefficient of the linear pressure drop* λ, *or:*

$$\lambda = \frac{2D_p}{\rho_p V_p^2 L_p} \Delta P_p = \frac{2D_m}{\rho_m V_m^2 L_m} \Delta P_m$$

With $V = \frac{4Q}{\pi D^2}$ *and* $\frac{D_m}{D_p} = 1$ *(same gallery) we obtain:*

$$\Delta P_p = \frac{\rho_p}{\rho_m} \frac{Q_p^2}{Q_m^2} \frac{L_p}{L_m} \Delta P_m = \frac{10^3}{1.25} \times \frac{7.5^2}{96.85^2} \times \frac{6,050}{500} \times 12 = 696.59 \; mbar$$

COMMENT 3.6.– *The hydroelectric gallery is a channel that brings water from a dam to a hydroelectric power station, flowing down a slight incline. As a general rule, the slope of the channel is between* 0.5/1,000 *and* 3/1,000.

In the majority of cases, the transfer of water is done under pressure (through a difference in level, the inclusion of a tank, or the need to obtain the maximum flow).

The forces involved are generally much greater in passages for hydroelectric developments.

Hydroelectricity, or hydroelectric power, utilizes the potential energy from flows of water (rivers, streams, waterfalls, sea currents, etc.). The kinetic energy of the water current is transformed into mechanical energy through a hydraulic turbine, and then into electrical energy using an alternator.

A hydroelectric plant is made up of a reservoir (taken "along the water flow" or in a dam) as well as an electrical production facility.

There are three main forms of hydroelectric power generation: pumping power transfer stations, marine power stations, and gravity supplied stations. The last of these, which we are studying here in this exercise, exist in three forms:

– power stations placed along a water flow, where the constant for emptying (or draining) is generally less than two hours;

– "guided" power stations, where the emptying constant is between two and two hundred hours;

– "lakes" (or reservoirs), where the emptying constant is more than two hundred hours.

There are a wide variety of hydroelectric developments depending on their geographical location, the type of water flow, the drop height, the nature of the dam, and its location relative to the power generation plant.

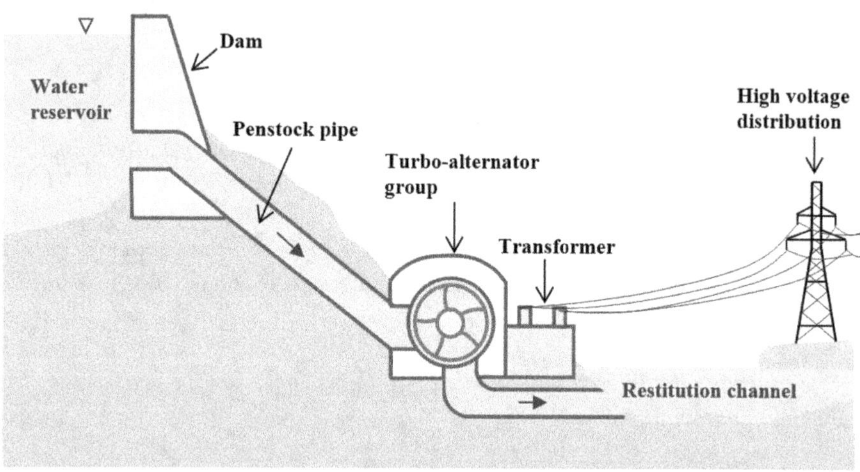

Figure 3.6. *Principle of operation of a gravity-powered hydroelectric power plant. For a color version of this figure, see www.iste.co.uk/sadchemloul/mechanics.zip*

3.2.10. *Exercise 10: movement of solid matter by a water current*

A completely submerged concrete block of a mass equal to $100\ kg$, of relative density 2.5 with relation to the water, is moved forward, sliding along the bottom of a river, when the velocity of the water current reaches $3\ m/s$.

What would the velocity of a water current need to be to move forward a geometrically similar block, with a mass of $150\ kg$ and a relative density 3.5 in relation to the water?

We will assume that the coefficient of friction f on the bottom of the river is the same in both cases, as well as the drag coefficient C_t.

REVIEW.– The frictional force exerted by the current on the block is given by $F_f = C_t \rho \frac{V^2}{2} S$, with ρ being the density of the fluid S in the frontal area (surface of the block seen from an infinite distance upstream) and V the velocity of the current.

3.2.10.1. *Solution*

The fully submerged concrete block is given in Figure 3.7.

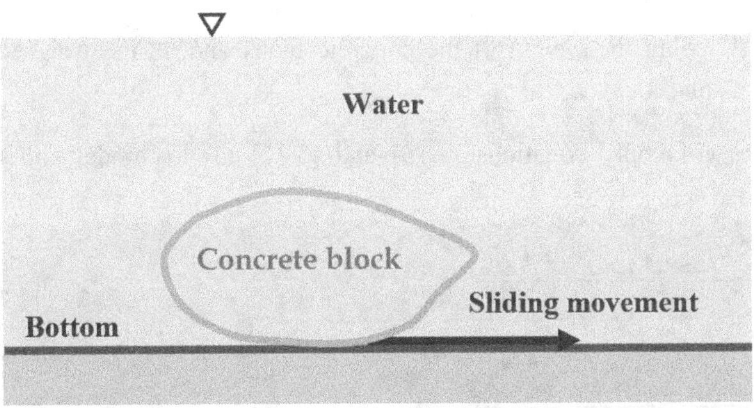

Figure 3.7. *Sliding a concrete block along the bottom of a river. For a color version of this figure, see www.iste.co.uk/sadchemloul/mechanics.zip*

The data of the prototype and of the model are shown in the table below:

Characteristics	Prototype	Model
Mass	$m_p = 100\ kg$	$m_m = 150\ kg$
Relative density of concrete block	$d_p = 2.5$	$d_m = 3.5$
Velocity flow of the water current	$V_p = 3\ m/s$	$V_m = ?$
Same fluid: water	$f_m = f_p$ and $C_{tp} = C_{tm}$	

Table 3.16. *Data of similarity characteristics: sliding of a concrete block along the bottom of a river*

The frictional force exerted by the current on the block is given by:

$$F_f = C_t \rho \frac{V^2}{2} S \qquad [3.16]$$

This force can also be determined by considering the coefficient of friction f between the concrete block and the bottom of the river:

$$F_f = f(P_b - P_e) \qquad [3.17]$$

with P_b being the weight of the concrete block and P_e the weight of the water displaced.

We will apply equations [3.16] and [3.17] to the model and to the prototype:

$$\frac{F_{f,m}}{F_{f,p}} = \frac{C_{t,m}}{C_{t,p}} \frac{\rho_m}{\rho_p} \frac{V_m^2}{V_p^2} \frac{S_m}{S_p} \qquad [3.18]$$

$$\frac{F_{f,m}}{F_{f,p}} = \frac{f_m(P_{b,m} - P_{e,m})}{f_p(P_{b,p} - P_{e,p})} \qquad [3.19]$$

Given that the similarity of the flows results in the drag and friction coefficients being equal for the model and the prototype ($C_{t,m} = C_{t,p}$ and

$f_m = f_p$) and that the same fluid is used ($\rho_m = \rho_p$), equations [3.18] and [3.19] are written as:

$$\frac{F_{f,m}}{F_{f,p}} = \frac{V_m^2}{V_p^2}\frac{S_m}{S_p} = \frac{V_m^2}{V_p^2}\frac{L_m^2}{L_p^2} \qquad [3.20]$$

$$\frac{F_{f,m}}{F_{f,p}} = \frac{(P_{b,m} - P_{e,m})}{(P_{b,p} - P_{e,p})} = \frac{(\rho_m - \rho_e)v_m}{(\rho_p - \rho_e)v_p} = \frac{(\rho_m - \rho_e)L_m^3}{(\rho_p - \rho_e)L_p^3} \qquad [3.21]$$

With v_m and v_p the volumes of the concrete block of the model and the prototype, respectively.

The equality between the equations [3.20] and [3.21] gives us:

$$\frac{V_m}{V_p} = \left(\frac{(\rho_m - \rho_e)L_m}{(\rho_p - \rho_e)L_p}\right)^{1/2} \qquad [3.22]$$

The ratio of the weights of the concrete blocks of the model and the prototype is:

$$\frac{P_{bm}}{P_{bp}} = \frac{\rho_m L_m^3}{\rho_p L_p^3} = \frac{100}{150} = \frac{2}{3}$$

The geometric scale ratio is:

$$\frac{L_m}{L_p} = \left(\frac{2}{3}\frac{\rho_p}{\rho_m}\right)^{1/3} = \left(\frac{2}{3}\frac{\rho_e d_p}{\rho_e d_m}\right)^{1/3} = \left(\frac{2}{3}\frac{d_p}{d_m}\right)^{1/3} = \left(\frac{2}{3}\times\frac{3.5}{2.5}\right)^{1/3} = 0.977$$

The ratio of powers is:

$$\frac{V_m}{V_p} = \left(\frac{(\rho_m - \rho_e)L_m}{(\rho_p - \rho_e)L_p}\right)^{1/2} = \left(\frac{d_m - 1}{d_p - 1}\frac{L_m}{L_p}\right)^{1/2} = \left(\frac{2.5-1}{3.5-1}\times 0.977\right)^{1/2} = 0.765$$

The velocity of the water current that will be able to move the prototype concrete block is:

$$V_p = \frac{V_m}{0.765} = \frac{3}{0.765} = 3.92\ m/s$$

COMMENT 3.7.– *The direct application of this exercise is the condition for sliding, which must always be included among the verifications to be carried out in the design of weighted structures in general, and marine structures in particular: docks, weights, and levees. One example is a dike, which is a longitudinal embankment, natural or artificial, most often made out of soil. The primary function of this structure is to prevent low lying lands along the levee from being submerged by the waters of a lake, river, or ocean. There are two categories that can be identified:*

– embankment levee: the height of this shell levee must be sufficient to prevent the breaking waves from flowing over the top, which reduces the effectiveness of the protection of lowlands on the other side. The dimensions of the levee are therefore determined based on the characteristics (essentially, the amplitude of the waves) generated by the strongest storm occurring at an average of once every 30 years. If a stronger storm occurs, the structure may be damaged to a greater or lesser extent, but would not be completely destroyed, and could be rehabilitated before the next storm. This kind of structure can be used, provided that the depth does not exceed 50 m;

– vertical levee: the wall consists of reinforced concrete blocks or caissons which, due to their own weight, resist the forces exerted on them by the waves: they must therefore be large in scale in order to be sufficiently heavy. When the wall is made of blocks stacked on top of each other, these blocks have a weight of up to one hundred tons, with this limit being set by the performance of the handling equipment used to set them up. Vertical levees are calculated for the height of the highest wave observed over 100 years. In order for the swell not to surge against the wall, it is necessary for the depth at the base of the wall to be two times the amplitude of this 100-year wave, and for the total depth at the base of the foundation to be 2.5–3 times this amplitude. It requires a height of at least 25 m for waves of 10 m.

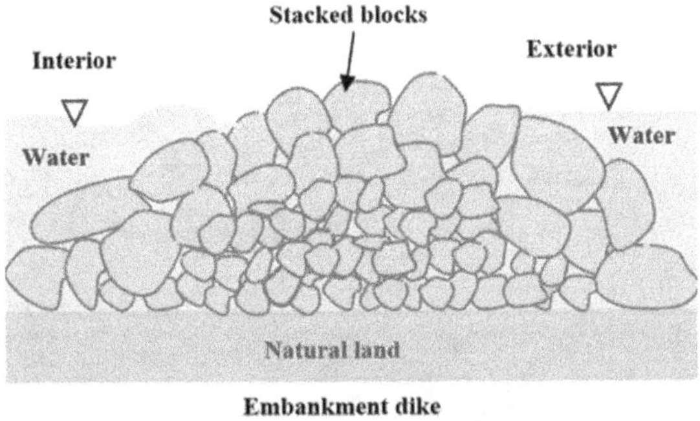

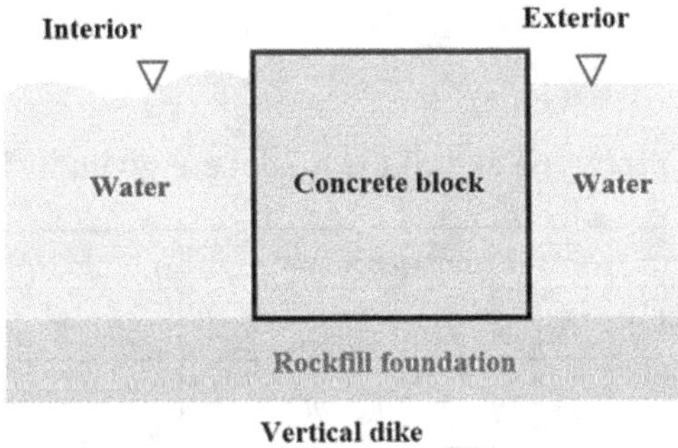

Figure 3.8. *The two main types of levees. For a color version of this figure, see www.iste.co.uk/sadchemloul/mechanics.zip*

3.2.11. *Exercise 11: a tapered body*

We would like to study the flow around a tapered body designed to be propelled underwater. We will create a scale model at 1:2, to be studied in a wind tunnel. Determine the speed at which the test should be performed if the expected speed in the water is $1.1\ m/s$.

We give: $\rho_{air} = 1.225\ kg/m^3$; $\rho_{water} = 10^3\ kg/m^3$; $\mu_{air} = 18 \times 10^{-6}\ PI$; $\mu_{water} = 0.0012\ kgm^{-1}s^{-1}$.

Will the test still be possible if the speed of the body in water is $9\ m/s$?

3.2.11.1. *Solution*

The following figure gives an example of a tapered body propelled under water.

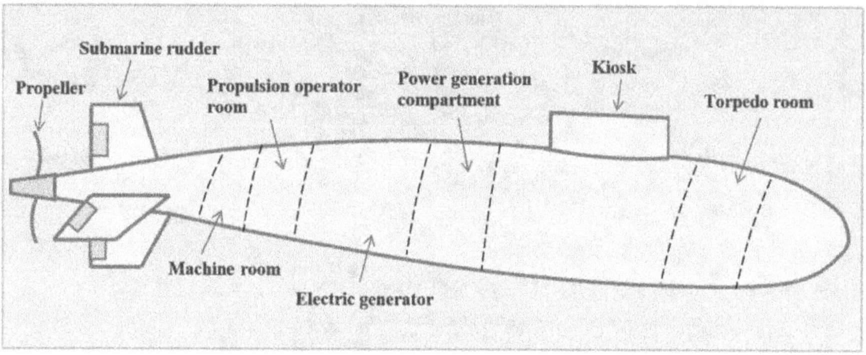

Figure 3.9. *A tapered body: example of a submarine*

The data of the prototype and of the model are given in Table 3.17:

Characteristics	Prototype: water	Model: air
Flow velocity of fluid	$V_p = ?$	$V_m = 1.1\ m/s$
Density	$\rho_p = 10^3 kg/m^3$	$\rho_m = 1.225\ kg/m^3$
Dynamic viscosity	$\mu_p = 0.0012\ kgm^{-1}s^{-1}$	$\mu_m = 18 \times 10^{-6} kgm^{-1}s^{-1}$
Geometric scale ratio	$\lambda_L = \frac{L_{R,m}}{L_{R,p}} = \frac{1}{2}$	

Table 3.17. *Data of similarity characteristics of a tapered body: example of a submarine*

NOTE 3.9.– *L_R represents a reference length of the tapered body: length, width, height, diameter, thickness, etc. This remark will be valid for all the exercises that follow.*

To determine the air velocity with which the test is to be performed, we must apply the Reynolds similarity condition:

$$Re_m = Re_p \text{ or } \frac{V_m L_{R,m}}{\nu_m} = \frac{V_p L_{R,p}}{\nu_p}$$

Using the similarity scales, we obtain $\lambda_V \lambda_L = \lambda_V$.

$$\lambda_V = \frac{\lambda_L}{\lambda_t} = \frac{12.24}{1/2} = 24.48 \text{ and } V_m = \lambda_V V_p = 24.48 \times 1.1 = 26.92\ m/s$$

If the speed of the body in the water is 9 *m/s*, then the velocity of the air in the wind tunnel is:

$$V_m = \lambda_V V_p = 24.48 \times 9 = 220.32\ m/s$$

NOTE 3.10.– *We can no longer consider air as a non-compressible fluid, thus the similarity is not valid. Generally, air is assumed to be a non-compressible fluid if its velocity is less than or equal to one third of the speed of sound (the speed of sound is* $340\ m/s$*).*

The resistance to forward movement is an important topic of naval hydrodynamics, since it is a question of determining the effort to be overcome in order to move a marine craft.

Historically, resistance to forward movement was one of the first topics studied in naval hydrodynamics, a discipline that arose in the eighteenth century. The works of Alembert[12]*, then Reech and Froude, allowed for the development of the theory of similarity applied to the problem of resistance to movement. This theory is currently what governs experiments on scale models. This domain (similarity) allows us to determine the performance of a submarine (or ship), because the determination of the resistance to forward movement gives us an initial idea of the power with which the craft must be equipped, which must be at least equal to the product of the resistance from the speed of movement.*

The propulsion devices (usually propellers) accelerate the flow around a submarine and tend to increase resistance due to the phenomenon of suction.

12 Jean le Rond d'Alembert, born on November 16, 1717, in Paris and who died in October 1783, was a mathematician, physicist, philosopher and encyclopedia writer.

3.2.12. *Exercise 12: model of a seaplane*

A scale model of a seaplane is made at $1:10$. It takes off at a speed of $50\ km/h$. Neglecting the influence of changes in the Reynolds number on the lift coefficient C_z, calculate the speed of takeoff of the prototype.

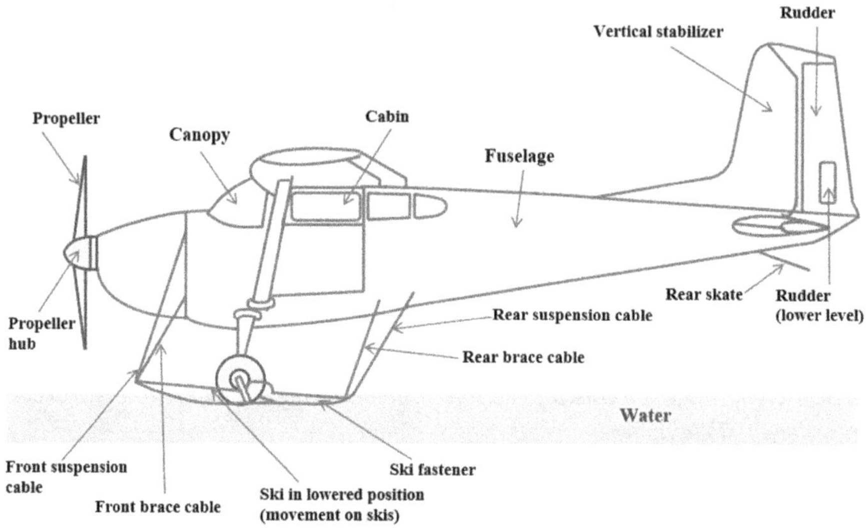

Figure 3.10. *Example of a model of a seaplane with floats*

3.2.12.1. *Solution*

The data of the prototype and of the model are shown in the table below:

Characteristics	Prototype: water	Model: water
Velocity of flow	$V_p = ?$	$V_m = 50\ km/h$
Geometric scale ratio	$\lambda_L = \frac{L_{R,m}}{L_{R,p}} = \frac{1}{10}$	

Table 3.18. *Data of the similarity characteristics of a seaplane*

The drag force is given by $F = C_z \rho \frac{V^2}{2} S$.

The seaplane takes off when the lift becomes equal to its weight.

The Reynolds number Re does not affect the coefficient of lift C_z, which does not vary, and since the same experimental conditions and the same fluid ($\rho_m = \rho_p$) are used here, we may write $C_{z,m} = C_{z,p}$.

The condition of dynamic similarity therefore gives us for the lift:

$$\frac{F_m}{F_p} = \frac{V_m^2}{V_p^2}\frac{S_m}{S_p}$$

Since it involves the same fluid ($\rho_m = \rho_p$) and the same position ($g_m = g_p$), the ratio of the weights of the model P_m and the prototype P_p is:

$$\frac{P_m}{P_p} = \frac{\rho_m}{\rho_p}\frac{v_m}{v_p}\frac{g_m}{g_p} = \left(\frac{L_{R,m}}{L_{R,p}}\right)^3$$

The relationship of the surface areas of the wings is $\frac{S_m}{S_p} = \left(\frac{L_{R,m}}{L_{R,p}}\right)^2$.

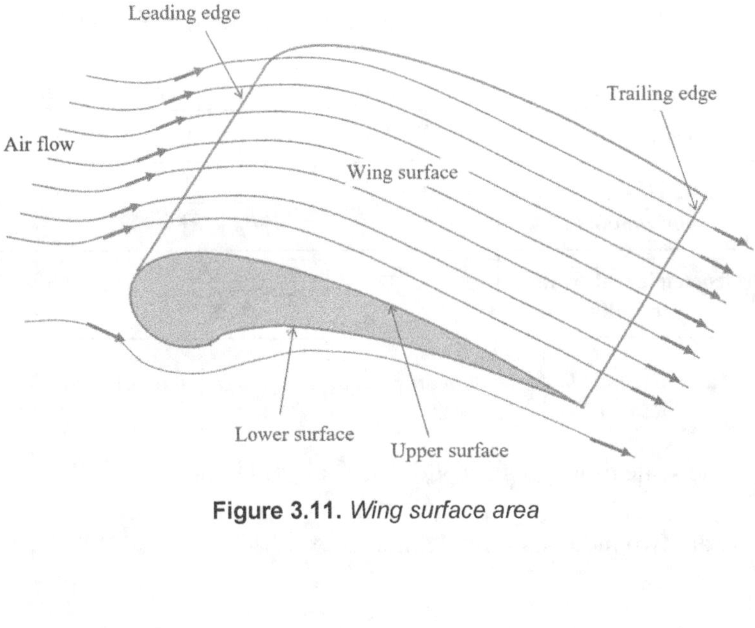

Figure 3.11. *Wing surface area*

The equality between the weight of the seaplane and the lift at takeoff gives us:

$$\frac{F_m}{F_p} = \frac{P_m}{P_p} \text{ or } \frac{V_m^2}{V_p^2} = \frac{L_{R,m}}{L_{R,p}}$$

The takeoff speed of the prototype is:

$$V_p = V_m \left(\frac{L_{R,p}}{L_{R,m}} \right)^{1/2} = 50 \times \sqrt{10} = 158.11 \; km/h$$

3.2.13. *Exercise 13: tide study*

A scale model intended for studying tides has been constructed at the scale of 1:500 horizontally and 1:80 in height. On this model, how long should the duration of a tide be, if its periodicity in nature is equal to 12 hours, 25 minutes?

Then, we have the same question for scales of 1:50,000 horizontally and 1:500 in height.

3.2.13.1. *Solutions*

The data of the prototype and of the model 1 are shown in the table below:

Characteristics	Prototype: water	Model 1: water
Duration of tide	$T_p = 12h\ 25mn$	$T_m = ?$
Geometric scale ratio horizontally	$\lambda_L = \frac{L_m}{L_p} = \frac{1}{500}$	
Geometric scale ratio vertically	$\lambda_h = \frac{h_m}{h_p} = \frac{1}{80}$	

Table 3.19. *Data of the similarity characteristics of a tide for model 1*

λ_L is the scale of lengths and λ_h is the scale of heights.

There are two methods for solving this exercise.

First method

Since this involves a free surface flow, the Froude similarity should be applied. Recall that the Froude number is the ratio of inertial forces and gravitational forces:

$$Fr_m = Fr_p \text{ or } \frac{V_m^2}{g_m h_m} = \frac{V_p^2}{g_g h_p}$$

Given that it is the same location ($g_m = g_p$), we obtain $\frac{V_m^2}{h_m} = \frac{V_p^2}{h_p}$, where h denotes the height of the tide.

The movement of the tide is mainly horizontal, and therefore the horizontal component of the velocity is dominant.

With L_h being the horizontal length crossed by the tide at the speed V for a time T, therefore, we can write for the last relationship:

$$\frac{L_{h,m}^2 / T_m^2}{h_m} = \frac{L_{h,p}^2 / T_p^2}{h_p}$$

The duration of the tide for the model is:

$$T_m = T_p \frac{L_{h,m}}{L_{h,p}} \left(\frac{h_p}{h_m} \right)^{1/2}$$

The duration of the tide is:

– for model 1:

$$T_m = (12h\ 25mn) \frac{1}{500} \sqrt{80} = 13mn\ 22s$$

– for model 2:

$$T_m = (12h\ 25mn) \frac{1}{50{,}000} \sqrt{500} = 20s$$

Second method

Instead of using the Froude similarity, we use the Strouhal similarity, which involves geometric, kinematic and timescale similarities. According to Bernoulli's equation for free surface flow, the squares of the velocity are proportional to the geometric heights of the tide.

With λ_T being the time scale and λ_V the speed scale, on the basis of Bernoulli's equation, we can write the relationship as $\lambda_V = \lambda_h^2$ without losing the generality.

On the basis of the Strouhal similarity, we obtain $\lambda_T = \frac{\lambda_L}{\lambda_V} = \frac{\lambda_L}{\lambda_h^{1/2}}$.

For model 1:

$$\lambda_T = \frac{\lambda_L}{\lambda_h^{1/2}} = \frac{\sqrt{80}}{\sqrt{500}} = 0.0179$$

And:

$$T_m = (12h\ 25mn)\frac{1}{500}\sqrt{80} = 13mn\ 22s$$

For model 2:

$$\lambda_T = \frac{\lambda_L}{\lambda_h^{1/2}} = \frac{\sqrt{500}}{\sqrt{50,000}} = 0.477\times 10^{-3}$$

And:

$$T_m = (12h\ 25mn)\frac{1}{50,000}\sqrt{500} = 20s$$

NOTE 3.11.– *Tides are the changes in ocean levels. These changes are due to two forces:*

– a force of gravitational attraction, exerted by a celestial body such as the moon or the sun;

– a centrifugal force, related to the rotation of the earth in its orbit, which is exerted identically at any point on the globe.

One of the primary goals of the study of tides is to research the relationship between the movement of the celestial bodies and the response of the oceans to the action of these gravitational forces, in order to establish formulas of prediction.

Tidal studies are also used in the fields of navigation, hydrography, port development and climate studies. The height of the tide at a time T is given by the following harmonic formula:

$$h(t) = z_o + \sum_i A_i(\cos q_i t + Vo_i + G_i)$$

with z_o *being the average level around which the water level oscillates,* A_i *the amplitude of the wave,* Vo_i *the values of astronomical arguments to* $t = 0$*, and* G_i *the orientation of elementary waves to Greenwich*[13]*, the angular velocity of the basic wave.*

The tide is measured using the tidal coefficient, which corresponds to the ratio, measured in Brest, France, of the tidal range[14] *occurring semi-diurnally*[15] *of the harmonic formula, divided by the mean tidal range for the spring tides occurring at the equinox. It is defined by the hydrographic and oceanographic service of the Navy. This service is a public establishment of a French administrative nature placed under the supervision of the Ministry of the Armed Forces.*

13 The Prime Meridian or Greenwich Meridian is the meridian (line of longitude) that serves as an international longitudinal reference. Greenwich is a neighborhood in London on the south bank of the Thames River.

14 The tidal range is the difference between a high tide and a low tide.

15 Semi-diurnal tidal cycle: this cycle occurs when there are two high tides and two low tides of roughly equal magnitude per day. This type of tide cycle is predominant in the Atlantic. A diurnal cycle is characterized by one high tide and one low tide per day. A semi-diurnal type of tide is unevenly diurnal: the tide always has two high and two low tides per day, but the heights of the high or subsequent low tides may be very different. With a mixed type of cycle, there are sometimes two high tides and two low tides per day (when the Moon is at the equator) and sometimes one high tide and one low tide per day (when the declination of the moon is close to its maximum amount).

3.2.14. *Exercise 14: transient gas flow*

In the laminar regime, the equation governing the transient gas flows in a duct with small cross-section can be written in the form:

$$\frac{\partial P}{\partial T} = C\frac{D^2}{64\mu}\frac{\partial^2 P^2}{\partial x^2}$$

where P is the pressure at time t and at the abscissa point x, D is the diameter of the duct, C is a constant depending on the shape of the section and μ the dynamic viscosity of the gas.

1) Find the relationship connecting the scales of similarity:

$$\lambda_L = \frac{x_1}{x_2},\ \alpha = \frac{P_1}{P_2},\ \theta = \frac{t_1}{t_2},\ \delta = \frac{D_1}{D_2}$$

2) Deduce what the influence would be of a decrease in length by one half and a multiplication of the diameter by 3 on the time taken to establish the same pressure with the same gas. What would this time become if it was initially equal to six months?

3.2.14.1. *Solutions*

1) To find the relationship between the different similarity scales, we will apply the equation $\frac{\partial P}{\partial T} = C\frac{D^2}{64\mu}\frac{\partial^2 P^2}{\partial x^2}$ to the two gas flows of a different nature:

– flow 1:

$$\frac{\partial P_1}{\partial T_1} = C_1\frac{D_1^2}{64\mu_1}\frac{\partial^2 P_1^2}{\partial x_1^2} \quad [3.23]$$

– flow 2:

$$\frac{\partial P_2}{\partial T_2} = C_2\frac{D_2^2}{64\mu_2}\frac{\partial^2 P_2^2}{\partial x_2^2} \quad [3.24]$$

Using the similarity scales:

$$\lambda_L = \frac{x_1}{x_2},\ \alpha = \frac{P_1}{P_2},\ \theta = \frac{t_1}{t_2},\ \delta = \frac{D_1}{D_2}$$

in equation [3.23], we obtain:

$$\frac{\alpha}{\theta}\frac{\partial P_2}{\partial T_2} = C_1 \frac{\alpha^2 \delta^2}{\lambda_L^2} \frac{D_2^2}{64\mu_1} \frac{\partial^2 P_2^2}{\partial x_2^2} \quad [3.25]$$

The ratio between equations [3.25] and [3.24] gives us:

$$\frac{\alpha}{\theta} = \frac{C_1}{C_2}\frac{\mu_1}{\mu_2}\frac{\alpha^2\delta^2}{\lambda_L^2}$$

or:

$$\theta = \frac{C_2}{C_1}\frac{\mu_2}{\mu_1}\frac{\lambda_L^2}{\alpha\delta^2} \quad [3.26]$$

NOTE 3.12.– *The constant C depends on the shape of the cross section, and since this uses the same duct, then* $C_1 = C_2$.

As the fluid is the same (a gas) $\mu_1 = \mu_2$, *equation* [3.26] *is therefore written as:*

$$\theta = \frac{\lambda_L}{\alpha\delta^2} \quad [3.27]$$

2) We will determine what the influence would be of a decrease in length by one half and a multiplication of the diameter by 3 on the time taken to establish the same pressure with the same gas.

The scales of similarity:

– decrease in length by half:

$$\lambda_L = \frac{x_1}{x_2} = \frac{1}{2}$$

– establish the same pressure:

$$\alpha = \frac{P_1}{P_2} = 1$$

– multiply the diameter by 3:

$$\delta = \frac{D_1}{D_2} = 3$$

The scale of time similarities is:

$$\theta = \frac{t_1}{t_2} = \frac{(1/2)^2}{1 \times 3^2} = \frac{1}{36}$$

Therefore, in order to establish the same pressure, the initial time will be divided by 36.

If the initial time is six months (180 days), then the time to achieve the same pressure will be $\frac{180}{36} = 5$ days, which demonstrates the interest of similarity.

3.2.15. *Exercise 15: model of a torpedo*

A torpedo model is tested in a test basin at a speed of $24\ m/s$. The prototype was designed to reach a speed of $6\ m/s$ at a temperature of $15°C$.

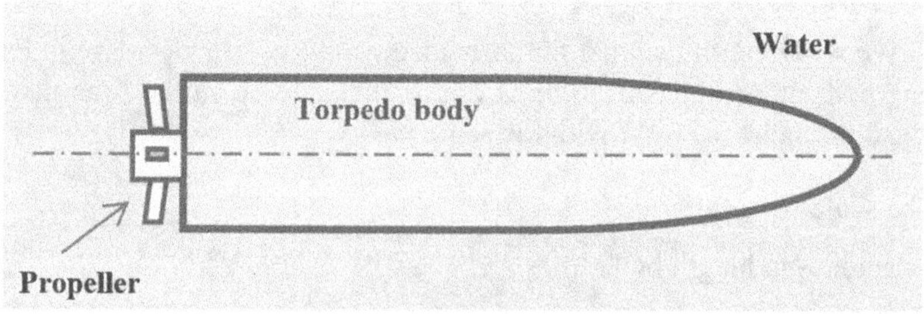

Figure 3.12. *Diagram of a torpedo*

1) What is the value of the reduction ratio (geometric scale ratio)?

2) What should the speed of the model be when tested in a wind tunnel at a pressure of 20 atmospheres and a temperature of $27°C$?

3.2.15.1. *Solutions*

The data of the model and of the prototype are shown in the table below:

Characteristics	Model	Prototype
Fluid	Water	Water
Speed of the torpedo	$V_m = 24\ m/s$	$V_p = 6\ m/s$
Temperature of the fluid	$T_p = T_m = 15°C$	
Geometric scale ratio	$\lambda_L = ?$	
Fluid	Air	Water
Temperature of the fluid	$T_m = 27°C$	$T_p = 15°C$
Pressure of the fluid	$P_m = 20\ bars$	$P_p = 1\ bar$
Speed of the torpedo	$V_m = ?$	$V_p = 6\ m/s$

Table 3.20. *Data of the similarity characteristics of a torpedo*

1) The reduction ratio shall be determined by applying the Reynolds similarity $Re_m = Re_p$, that is:

$$\frac{V_m D_m}{\nu_m} = \frac{V_p D_p}{\nu_p} \text{ or } \frac{V_m}{V_p}\frac{D_m}{D_p} = \frac{\nu_m}{\nu_p}$$

Using the similarity scales $\lambda_V \lambda_L = \lambda_\nu$ and keeping in mind that it is the same fluid ($\lambda_\nu = 1$), we obtain:

$$\lambda_L = \frac{\lambda_\nu}{\lambda_V} = \frac{1}{\lambda_V} = \frac{6}{24} = \frac{1}{4}$$

2) For the speed of the torpedo in the wind tunnel, the Reynolds similarity relationship $\lambda_L \lambda_V = \lambda_\nu$ is used.

$$\lambda_\nu = \frac{\nu_{air}(27\,°C)}{\nu_{water}(15\,°C)} = \frac{\nu_m(27\,°C)}{\nu_p(15\,°C)}$$

From the table of the thermo-physical properties of air, we obtain:

$$\mu_m(27\ °C) = \mu_p(27\ °C) = 1.857 \times 10^{-5}\, kgm^{-1}s^{-1}$$

The density ρ_{air} of air is determined from the specific weight ϖ_{air}, which in turn is determined from Boyle's law and Mariotte's law[16].

$$\rho_m = \rho_{air} = \frac{\varpi_m(P,T)}{g},\ \text{with: } \varpi_m = \varpi_{air} = \frac{P}{RT}$$

With $P_m = 20\ bars$, $T_m = 27 + 273 = 300\ °K$, $R = 29{,}3\ m/°K$ and $g = 10\ m/s^2$, we obtain $\varpi_m = 227.53\ N/m^3$.

$$\nu_m(T = 27\ °C, P = 20\ bars) = \frac{\mu_m(T = 27\ °C, P = 20\ bars)}{\rho_m(T = 27\ °C, P = 20\ bars)}$$

$$\nu_m(T = 27\ °C, P = 20\ bars) = \frac{1.857 \times 10^{-5}}{22.753} = 0.0816 \times 10^{-6}\ m^2/s$$

From the table of the thermo-physical properties of water, we obtain:

$$\nu_p(T = 15\ °C) = \nu_{water}(T = 15\ °C) = 1.142 \times 10^{-6}\ m^2/s$$

The scale ratio of kinematic viscosity:

$$\lambda_\nu = \frac{\nu_m}{\nu_p} = \frac{0.0816 \times 10^{-5}}{1.142 \times 10^{-6}} = 0.714$$

The scale ratio of speed:

$$\lambda_V = \frac{\lambda_\nu}{\lambda_L} = \frac{0.714}{1/4} = 2.856$$

16 Robert Boyle, born in Waterford (Ireland) on January 25, 1627, and who died on December 30, 1691, was a physicist and chemist. Edme Mariotte, born ca. 1620 in Dijon (France), and who died on May 12, 1684, was a physicist and botanist.

The speed of the torpedo in a wind tunnel at a pressure of 20 atmospheres and a temperature of $27°C$ is:

$$V_m = \lambda_V V_p = 2.856 \times 6 = 17.13 \text{ m} / s$$

From the table of the thermo-physical properties of water, we obtain:

$$\nu_p(T = 27\ °C) = \nu_{water}(T = 27\ °C) = 8.568 \times 10^{-7}\ m^2 / s$$

The scale ratio of kinematic viscosity:

$$\lambda_\nu = \frac{\nu_m}{\nu_p} = \frac{0.0816 \times 10^{-5}}{8.568 \times 10^{-7}} = 0.9523$$

The scale ratio of speed:

$$\lambda_V = \frac{\lambda_\nu}{\lambda_L} = \frac{0.9523}{1/4} = 3.8092$$

The speed of the model at a pressure of 20 atmospheres and a temperature of $27°C$ is:

$$V_m = \lambda_V V_p = 3.8092 \times 6 = 22.85\ m / s$$

NOTE 3.13.– *A torpedo is a self-propelled projectile that moves through water and is intended to destroy ships or submarines. It can be launched from a ship, submarine, maritime patrol airplane, helicopter, or carried by a missile.*

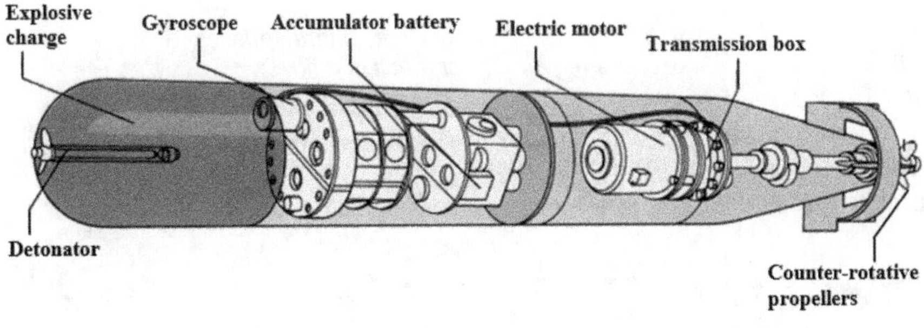

Figure 3.13. *Example of a torpedo. For a color version of this figure, see www.iste.co.uk/sadchemloul/mechanics.zip*

3.2.16. *Exercise 16: movement of a ball in a fluid*

We want to estimate the frictional force on a prototype ball of diameter $d_p = 0.3\ m$ moving in water ($\rho_{water} = 10^3\ kg/m^3; \nu_{water} = 1.307 \times 10^{-6} m^2/s$) with a speed of $V_p = 2.5\ m/s$.

1) What should the air velocity ($\rho_{air} = 1.2\ kg/m^3$; $\nu_{air} = 1.5 \times 10^{-5} m^2/s$) be in a tunnel in which a model ball of diameter $d_m = 0.15\ m$, similar to the prototype ball, is placed?

2) Determine the friction force F_p of water on this prototype ball if the friction force of air on the model ball is $F_m = 25\ N$.

3.2.16.1. *Solutions*

The data of the model ball and of the prototype ball are shown in the table below:

Characteristics	Model ball	Prototype ball
Fluid	Air	Water
Density	$\rho_m = 1.2\ kg/m^3$	$\rho_p = 10^3\ kg/m^3$
Dynamic viscosity	$\nu_m = 1.5 \times 10^{-5} m^2/s$	$\nu_p = 1.307 \times 10^{-6} m^2/s$
Diameter of the ball	$d_m = 0.15\ m$	$d_p = 0.3\ m$
Speed of the ball	$V_m = ?$	$V_p = 2.5\ m/s$
Friction force	$F_m = 25\ N$	$F_p = ?$

Table 3.21. *Data of the similarity characteristics of the movement of a ball in a fluid*

1) With the flows around the two balls being in similarity, their Reynolds numbers are equal, that is:

$$Re_m = Re_p \text{ or } \frac{V_m D_m}{V_p D_p} = \frac{\nu_m}{\nu_p}$$

Thus, the speed of the model is:

$$V_m = \frac{\nu_m}{\nu_p}\frac{D_p}{D_m}V_p = \frac{1.5\times10^{-5}}{1.307\times10^{-6}}\times\frac{0.30}{0.15}\times2.5 = 57.4\ m/s$$

2) The frictional force is given by:

$$F = C_D\rho\frac{V^2}{2}S = C_D\rho\frac{V^2}{2}\frac{\pi D^2}{4}$$

C_D is the drag coefficient that depends on the Reynolds number *Re*, $C_D = f(Re)$.

The similarity of the flows of water and air around the two balls (model and prototype), is achieved when the two flows (model and prototype) have the same Reynolds number, and therefore the same drag coefficient, which allows us to write:

$$f(Re_m) = f(Re_p) \text{ or } C_{D,m} = C_{D,p}$$

that is:

$$\frac{F_m}{F_p} = \frac{\rho_m}{\rho_p}\frac{V_m^2}{V_p^2}\frac{D_m^2}{D_p^2}$$

The force exerted by the air on the prototype ball is:

$$F_p = \frac{\rho_p}{\rho_m}\frac{V_p^2}{V_m^2}\frac{D_p^2}{D_m^2}F_m = \frac{10^3}{1.25}\times\frac{2.5^2}{57.4^2}\times\frac{0.30^2}{0.15^2}\times25 = 158.08\ N$$

NOTE 3.14.– *The flow generated by the uniform rectilinear movement of a sphere of radius R and velocity V in a fluid is equivalent to that of a flow around a stationary sphere. The velocity of the fluid at a long distance upstream and downstream of the sphere is therefore V.*

Practical experience shows that the sphere reaches a constant limit speed. This suggests the existence of a force other than the weight and Archimedian buoyancy, and associated with the friction of the fluid on the sphere (viscosity effect). The fluid exerts, on the sphere of diameter $d = 2R$, a force F in the direction of the flow, known as drag.

Practical experience shows that the progression of the drag force with the velocity of the fluid is universal (independent of the fluid and the constituent material of the sphere) if expressed in the form $C_D = f(Re)$ *(see Figure 2.12).*

The definition of the drag coefficient is:

$$C_D = \frac{F}{\rho \frac{V^2}{2} S} = \frac{\tau_p}{\rho \frac{V^2}{2}}$$

with τ_p *the shear stress along the wall,* S *the front surface of the obstacle known as "frontal area", here given as* $S = \pi R^2$.

The shape of the curve depends on the geometric form of the object, but all drag curves have the same general contour. The following limiting behaviors can be identified:

– for small Reynolds numbers $Re < 1$*, the regime is laminar and the drag coefficient* C_D *is inversely proportional to* Re*. It is described by the relationship* $C_D = \frac{24}{Re}$*, and the expression of the drag is represented by the Stokes formula* $F = -\ 6\pi\mu RV$ *where* μ *is the dynamic viscosity of the fluid;*

– for larger Reynolds numbers $10^3 < Re < 10^6$*,* C_D *is constant, which reflects the stabilization of drag. We thus have a quadratic dependence on the speed;*

– for $Re \approx 10^6$*, the drag drops sharply; this is known as a drag crisis;*

– in common industrial problems, the Reynolds criterion is always between 10^{-4} *and* 4×10^5 *in such a way that an attempt has been made to interpret the variation of the Newton criterion using a unique relationship of the type* $C_D = aRe^n$.

The values of the coefficients a *and* n *are as a function of the Reynolds number according to the table below:*

Flow regime of	Reynolds number	a	n
Stokes	$10^{-4} < Re < 1$	24	1
Allen	$1 < Re < 10^3$	18.5	0.6
Newton	$10^3 < Re < 4\times10^5$	0.44	0

Table 3.22. *Values of constants a and n as a function of the Reynolds number*

3.2.17. *Exercise 17: similarity of the movement of an airship*

It is desired to set an airship in motion of length L_p and speed $V_p = 6\ m/s$ in the air ($\rho_{air} = 1.205\ kg/m^3$; $\mu_{air} = 1.8\times10^{-5}\ kgm^{-1}s^{-1}$). A model of this prototype of length $L_m = L_p/30$ is set in motion in water, of density $\rho_{water} = 998\ kg/m^3$, and dynamic viscosity $\mu_{water} = 10^{-3}\ kgm^{-1}s^{-1}$.

1) Determine the speed of the model.

2) The friction force measured on the model is $F_m = 2,700N$. Calculate the friction force F_p estimated on the prototype.

3) Calculate the power p_p required in kW to propel the prototype.

3.2.17.1. *Solutions*

The data of the model and of the prototype of airship are shown in the table below:

Characteristics	Model: water	Prototype: air
Speed	$V_m = ?$	$V_p = 6\ m/s$
Geometric scale ratio	$\frac{L_m}{L_m} = \frac{1}{30}$	
Density	$\rho_m = 998\ kg/m^3$	$\rho_p = 1.205\ kg/m^3$
Dynamic viscosity	$\mu_m = 10^{-3}\ kgm^{-1}s^{-1}$	$\mu_p = 1.8\times10^{-5}\ kgm^{-1}s^{-1}$

Table 3.23. *Data on the similarity of a moving airship*

An example of the airship is given in the figure below:

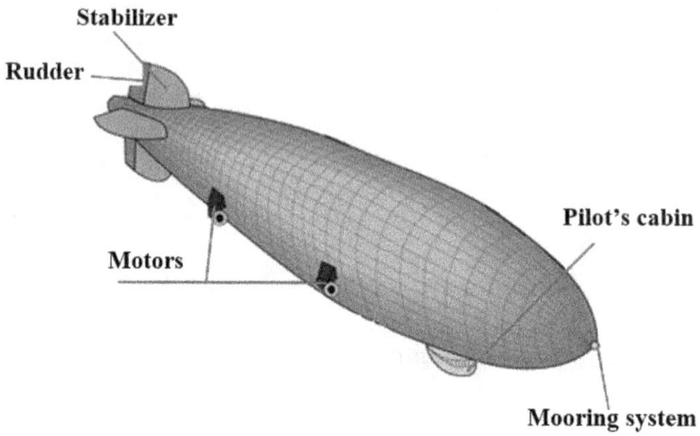

Figure 3.14. *Example of an airship*

1) To determine the speed of the model V_m, we will apply the Reynolds similarity:

$$Re_m = Re_p \text{ or } \frac{\rho_m V_m D_m}{\mu_m} = \frac{\rho_p V_p D_p}{\mu_p}$$

And therefore:

$$V_m = \frac{\rho_p}{\rho_m} \frac{D_p}{D_m} \frac{\mu_m}{\mu_p} V_p = \frac{1.205}{998} \times \frac{30}{1} \times \frac{10^{-3}}{1.8 \times 10^{-5}} \times 6 = 12.07 \text{ m/s}$$

2) The frictional force of a fluid on a body with surface area S is:

$$F = \tau_p S$$

With $\tau_p = C_D \rho \frac{V^2}{2}$ the wall shear stress, and C_D the coefficient of friction or drag.

Due to the similarity, the coefficients of friction of the prototype $C_{D,p}$ and the model $C_{D,m}$ are equal ($C_{D,m} = C_{D,p}$), that is:

$$\frac{\tau_m}{\tau_p} = \frac{\rho_m}{\rho_p}\frac{V_m^2}{V_p^2}$$

Taking into account the definition of the stress at the wall, we may write:

$$\frac{F_m / S_m}{F_p / S_p} = \frac{\rho_m}{\rho_p}\frac{V_m^2}{V_p^2}$$

The friction force exerted on the prototype is:

$$F_p = \frac{\rho_p}{\rho_m}\frac{V_p^2}{V_m^2}\frac{S_p}{S_m}F_m = \frac{\rho_p}{\rho_m}\frac{V_p^2}{V_m^2}\left(\frac{L_p}{L_m}\right)^2 F_m$$

$$F_p = \frac{1.205}{998}\times\left(\frac{6}{12.07}\right)^2\times\left(\frac{30}{1}\right)^2\times 2,700 = 725\ N$$

3) The power p_p required to power the prototype is:

$$p_p = F_p V_p = 725\times 6 = 4.35\ kW$$

COMMENT 3.8.– *The airship was one of the first types of aircraft invented by man. Today, it has attracted a certain level of interest, with the potential to improve its effectiveness as a means of transport. It was created before all other types of aircraft, and has seen use in civil and military aviation.*

The airship holds its altitude because of the principle of Archimedian buoyancy. This refers to the force exerted on a body by the liquid or gas in which it is immersed. The force is exerted vertically, upward from below (in the opposite direction of gravitational attraction). It is equal to the attraction exerted on a volume of the gas or liquid, equivalent to the immersed volume of the body in question.

The terrestrial attraction exerted on the airship is equal to the buoyancy generated by the atmosphere, which maintains its altitude constant. A gas with a lower density than air (hydrogen, helium, or hot air) compensates for the weight of the structure, nacelle[17] *and engines. Thus, the total density of the airship is equal to that of the air. To gain altitude, the weight (either water, sand, or air) contained in the reserve areas provided for this purpose is released, and to lose altitude, part of the flotation gas is released.*

3.2.18. *Exercise 18: resistance to the movement of a ship*

The purpose of this problem is to determine the resistance exerted against the forward movement of a ship.

1) What quantities determine the resistance force to the advancement F_a of a vessel with a given hull geometry?

2) Taking as repeating variables the physical quantities: the length L_n of the ship, the forward speed V_a of the ship, and the density of the water ρ, determine the expression for the force of resistance to the forward movement F_a.

3) To establish the law governing the resistance to the advancement of the ship, a 1:10 scale model is to be used. Write the corresponding similarity conditions. Deduce the speed at which the model needs to move.

In the experiment carried out on the model in the ship test basin, the fluid used must be water. Can dynamic similarity be complete?

3.2.18.1. *Solutions*

1) The relationship of the resistance to the advancement F_a of the boat depends on the density ρ and dynamic viscosity μ of the water, the length L_n and speed of the ship V_a, and the acceleration from gravity g, that is:

$$F_a = f(\rho, \mu, L_n, V_a, g)$$

17 The nacelle (made from wicker) is suspended from a hot air balloon, and is the area where the balloon operators ride. The hot air balloon is a lighter-than-air craft, composed of a basket suspended from a lightweight air compartment, which takes flight as a result of the heated air inside, subjected to the force of Archimedian buoyancy. The hot air balloon was invented by the French-born Montgolfier brothers, Joseph-Michel and Jacques-Étienne, in 1782.

The number of physical quantities or physical variables is therefore $n = 6$.

NOTES 3.15.–

– The acceleration from gravity g has been taken into account in this physical phenomenon, because it is a factor in phenomena such as waves, wakes, and in general, the deformation of the fluid's surface, created by the movement of the boat.

– It is assumed that the depth at which the boat floats in the water does not vary with its speed (there is no hydroplaning).

2) The dimension of the physical quantities is:

$$[F_a] = MLT^{-2},\ [\rho] = ML^{-3},\ [\mu] = ML^{-1}T^{-1}$$

$$[L_n] = L,\ [V_a] = LT^{-1},\ [g] = LT^{-1}$$

The number of the basic dimension is $k = 3$, or three terms in π.

Considering L_n, V_a, and ρ as repeating quantities, the terms π are written as:

$$\begin{cases} \pi_1 = L_n^{\alpha} V_a^{\beta} \rho^{\gamma} F_a \\ \pi_2 = L_n^{\alpha} V_a^{\beta} \rho^{\gamma} \mu \\ \pi_3 = L_n^{\alpha} V_a^{\beta} \rho^{\gamma} g \end{cases}$$

Using the dimension of the terms in π, we determine:

– term π_1:

$$\begin{cases} M: \gamma + 1 = 0 \\ L: \alpha + \beta - 3\gamma + 1 = 0 \\ T: \beta + 2 = 0 \end{cases} \Rightarrow \begin{cases} \alpha = -2 \\ \beta = -2 \\ \gamma = -1 \end{cases},\ \pi_1 = \frac{F_a}{\rho V_a^2 L_n^2}$$

– term π_2:

$$\pi_2 = \frac{\mu}{\rho V_a L_n}$$

which is the inverse of the Reynolds number, but dimensional analysis allows us to write $\pi_2 = Re$;

– term π_3:

$$\pi_3 = \frac{gL}{V_a^2} = \frac{1}{F_r^2}$$

which is the inverse of the square of Froude's number.

The three terms in π confirm the following relationship:

$$F(\pi_1, \pi_2, \pi_3) = 0$$

The resistance to movement is expressed as:

$$F_a = \rho V_a^2 L_n^2 F_1(Re, Fr)$$

The resistance to movement thus depends on the Reynolds number, the Froude number, the term L_n^2 representing the front surface of the obstacle known as "Master torque", and the dynamic pressure ρV_a^2.

3) Thus, complete similarity between the model ship and the prototype ship will be reached when, in addition to the geometric similarity, it is necessary to implement mechanical similarity (Reynolds similarity and Froude similarity). These last two conditions of similarity allow us to write:

$$\begin{cases} Re_m = Re_p \\ Fr_m = Fr_p \end{cases} \Rightarrow \begin{cases} \dfrac{\rho_m V_{a,m} L_{n,m}}{\mu_m} = \dfrac{\rho_p V_{a,p} L_{n,p}}{\mu_p} \\ \dfrac{V_{a,m}}{\sqrt{g_m L_{n,m}}} = \dfrac{V_{a,p}}{\sqrt{g_p L_{n,p}}} \end{cases}$$

Given that the experiment is carried out on the same location ($g_m / g_p = 1$), the ratio of speeds obtained from the Froude similarity condition gives us:

$$\frac{V_{a,m}}{V_{a,p}} = \left(\frac{g_m}{g_p}\right)^{1/2} \left(\frac{L_{n,m}}{L_{n,p}}\right)^{1/2} = \frac{1}{\sqrt{10}}$$

Taking this last result into account, the Reynolds similarity condition gives us:

$$\frac{\rho_m}{\rho_p} \frac{V_{a,m}}{V_{a,p}} \frac{L_{n,m}}{L_{n,p}} \frac{\mu_p}{\mu_m} = 1$$

With $\frac{L_{n,m}}{L_{n,p}} = \frac{1}{10}$ we obtain:

$$\frac{\rho_m}{\rho_p} \frac{\mu_p}{\mu_m} = 10 \times \sqrt{10}$$

NOTE 3.16.– *It would therefore be necessary for the model to move three times faster than the real ship, but it would also be necessary to find a fluid that is at the same time much denser and much less viscous than water. Such a fluid does not exist! Therefore, it is not possible to achieve complete similarity.*

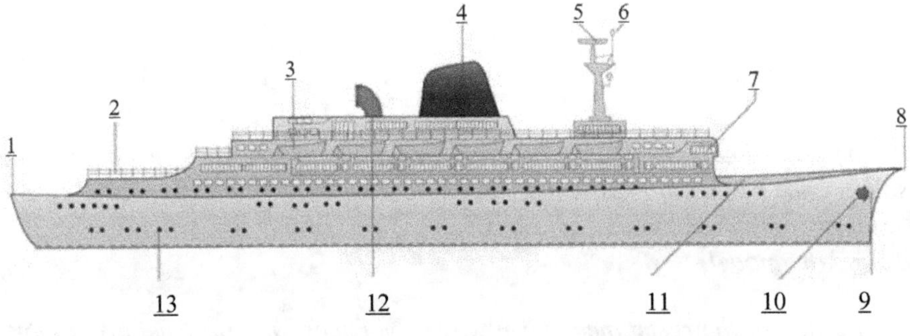

Figure 3.15. *Example of a ship designed for carrying passengers: 1) pump; 2) deck; 3) lifeboat; 4) chimney; 5) radar; 6) masthead light; 7) bridge; 8) bow; 9) anchor; 10) study; 11) bulwark; 12) windsock; 13) porthole. For a color version of this figure, see www.iste.co.uk/sadchemloul/mechanics.zip*

COMMENT 3.9.– *The resistance to movement* F_a *can be broken down into two independent forces:*

– a force F_{af} *linked to the viscosity of the fluid and reflecting the friction effect of the water along the hull due to the wet surface of the boat, that of the waves, the resistance of vortices, that of protrusions (rudders, fins, anti-roll keels, thin planes, shaft lines, etc.). In this case,* $F_1(Re, Fr) = F_2(Re)$ *represents the coefficient of friction* C_f *and the friction force is given by* $F_{af} = C_f \rho \frac{V_a^2}{2} S$ *with:*

$$C_f = \frac{0.074}{(Re)^{0.2}} \text{ for } Re < 10^7 \text{ (Prandtl relationship)}$$

$$C_f = \frac{0.455}{(\log_{10} Re)^{2.584}} \text{ for } Re > 10^7 \text{ (Prandtl-Schlichting relationship)}^{18}$$

– a force F_{ao} *representing the wave resistance (aerodynamic resistance), that is, the resistance opposed by the series of waves created by the boat on the surface of the water. In that case* $F_1(Re, Fr) = F_3(Fr)$.

Thus, the resistance to movement F_a *can be written as:*

$$F_a = \rho V_a^2 L_n^2 \left(F_2(Re) + F_3(Fr)\right)$$

The experiment provides the total value of the resistance around the model, and the friction resistance will be calculated. The difference between these two resistors makes it possible to determine the wave resistance around the model.

The sum of the resistances from waves, vortices, and protrusions is more commonly referred to as "residual resistance".

18 Hermann Schlichting, born on September 22, 1907, and who died on June 15, 1982, was a German fluid dynamics engineer.

Friction resistance arises from the adhesion of a thin film of water, adhering to the surface of the hull and moving at the same speed as the boat. However, at a certain distance from the surface of the hull, the water is stationary. Between these two extremes, the movement of the water particles exhibits complex behaviors that generate forces as a result of the difference in speed within this zone. The thickness of this boundary layer varies along the length of the boat until it breaks at a point of separation. Then, a transient zone is formed in which the water particles require more space to move, until finally becoming totally disconnected, and therefore turbulent. The water molecules are shaken and are driven in all directions. Thus, whirlpools (vortices) form behind the separation point.

To carry out the tests, an exact scale model must be built. An effort is made to make models large enough to minimize measurement errors, to be able to apply an extrapolation to the full scale. The model is connected to a tensile tube, which itself is attached to the platform running on a rail, just like an electric train. This traction tube is connected to a three-dimensional balance, which is capable of measuring forces in each direction.

3.2.19. *Exercise 19: mixing tank*

A laboratory tank with a volume $v_c = 5\ liters$ equipped with a counter-blades and a Rushton turbine of diameter $D_t = 9.2\ cm$ has given a satisfactory performance for stirring power $P_t = 5\ W$ in turbulent regime.

1) Give the expression of the power dissipated by the turbine in the fluid, which will depend on the speed of rotation N_t of the turbine, the diameter D_t of the turbine, the density ρ and viscosity μ of the fluid, and the gravitational acceleration g.

2) What should be the turbine diameter be for a tank of volume $v_c = 5\ m^3$?

3) We wish to extrapolate this reactor through maintaining the stirring speed $N_t = 3{,}000\ rpm$; would this be reasonable?

4) What should be the speed and stirring power, depending on whether the power per unit volume or the speed at the end of the blade is maintained?

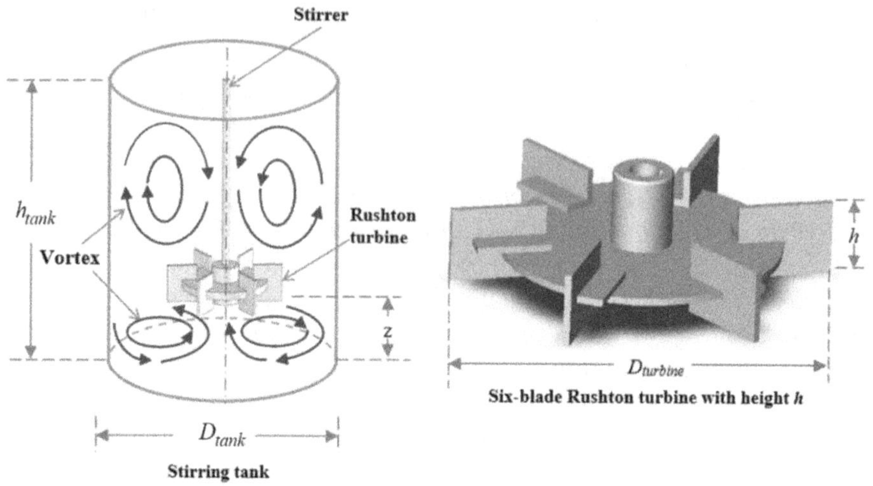

Figure 3.16. *Stirring tank with six-blade Rushton turbine*[19]. *For a color version of this figure, see www.iste.co.uk/sadchemloul/mechanics.zip*

NOTE 3.17.– *This is a standard Rushton turbine with six blades. The turbine of diameter* $D_{turbine} = 0.33D_{\text{tank}}$ *is placed in a cylindrical tank with no baffles at an axial position equal to* $z = 0.33H_{\text{tank}}$. *The height of the tank is equal to its diameter (* $H_{\text{tank}} = D_{\text{tank}}$ *).*

3.2.19.1. *Solutions*

1) In order to set a liquid in motion for mixing with another liquid, whether miscible or not, or for the creation of a solid suspension, or to cause the dispersion of a gaseous phase, energy is needed. This is provided by the agitator (or by the turbine) and has the relationship of:

$$p_t = f(N_t, D_t, \rho, \mu, g)$$

The number of physical quantities or variables is $n = 6$ and their dimensions are:

19 John Henry Rushton, born on November 25, 1905, in New London (United States) and who died on June 16, 1985, was an engineer, chemist, university professor and top-level expert on processes of mixing and transfers of mass.

$$[p_t] = ML^2T^{-2}, [N_t] = T^{-1}, [D_t] = L$$

$$[\rho] = ML^{-3}, [\mu] = ML^{-1}T^{-1}, [g] = LT^{-2}$$

The number of fundamental quantities or basic dimensions is $k = 3$.

The number of terms in π is $n - k = 3$.

Considering N_t, D_t, and ρ as repeating variables, the terms π are written as:

$$\begin{cases} \pi_1 = N_t^{\alpha} D_t^{\beta} \rho^{\gamma} p_t \\ \pi_2 = N_t^{\alpha} D_t^{\beta} \rho^{\gamma} g \\ \pi_3 = N_t^{\alpha} D_t^{\beta} \rho^{\gamma} \mu \end{cases}$$

Using the dimension of the terms in π, we determine:

$$\pi_1 = \frac{p_t}{\rho N_t^3 D_t^5}$$

This term represents the power number:

$$N_p = \frac{p_t}{\rho N_t^3 D_t^5}$$

$$\pi_2 = \frac{g}{N_t^2 D_t}$$

This term is proportional to the Froude number, effectively:

$$\pi_2 = \frac{g}{N_t^2 D_t} = \frac{gD_t}{N_t^2 D_t^2} = \frac{gD_t}{U_t^2} = \frac{1}{Fr^2}$$

With U_t being the linear or peripheral speed of the turbine.

$$\pi_3 = \frac{\mu}{\rho N_t D_t^2}$$

This term is the inverse of the Reynolds number, in effect:

$$\pi_3 = \frac{\mu}{\rho N_t D_t^2} = \frac{\mu}{\rho N_t D_t D_t} = \frac{\mu}{\rho U_t D_t} = \frac{1}{Re}$$

With U_t being the linear or peripheral speed of the turbine.

The terms in π confirm the relationship $F(\pi_1, \pi_2, \pi_3) = 0$.

The power dissipated by the turbine in the fluid is:

$$p_t = \rho N_t^3 D_t^5 F_1(Fr, Re)$$

NOTE 3.18.– *This power can be expressed using the number of powers* $N_p = F_1(Fr, Re)$.

Each stirrer, in addition to the number of powers, is characterized by:

– the pumping number N_{op} *(ability to pump out the liquid);*

– the circulation number N_{oc} *(the ability to move the liquid inside the tank).*

It is necessary to know these dimensional numbers in order to properly scale and design the stirring and mixing processes.

2) The data of the similarity between the flows in the two stirrers are:

Characteristics	Tank 1	Tank 2
Volume of the tank	$v_{c,1} = 5$ *litres*	$v_{c,2} = 5\ m^3$
Diameter of the turbine (or stirrer)	$D_{t,1} = 9\ cm$	$D_{t,2} = ?$
Power of the stirrer	$p_{t,1} = 5$ W	$p_{t,2} = ?$
Same fluid	$\rho_1 = \rho_2$, $\mu_1 = \mu_2$	

Table 3.24. *Data of the similarity of the flow in a mixing tank*

REVIEW.– Considering the power number $N_p = \dfrac{p_t}{\rho N_t^3 D_t^5}$ and knowing that the volume of the tank is given as $v_c \propto D_t^3$, and that the same fluid is being used ($\rho_1 = \rho_2$), we can write $\dfrac{p_t}{v_c} \propto \rho N_p N_t^3 D_t^2$.

The conservation of power per unit volume applied to both volumes, $v_{c,1} = 5\ liters$ and $v_{c,2} = 5\ m^3$, gives $N_{p,1} N_{t,1}^3 D_{t,1}^2 = N_{p,2} N_{t,2}^3 D_{t,2}^2$.

In the laminar regime ($Re < 10$), the number of powers is given by:

$$N_p = \frac{1}{Re}$$

that is $N_p = \dfrac{1}{Re} \propto \dfrac{1}{N_t D_t^2}$:

Since the same tank ($D_{t,1} = D_{t,2}$) is used, the application of the second of these relationships to both volumes gives us $N_{t,1} = N_{t,2}$.

The transition regime between the laminar regime and turbulent regime is such that $1 < Re < 10^4$.

In a turbulent regime ($Re > 10^4$), since N_p is constant ($N_{p,1} = N_{p,2}$), therefore:

$$N_{t,1}^3 D_{t,1}^2 = N_{t,2}^3 D_{t,2}^2$$

The vortex effect only appears for $Re > 10^3$.

The size ratio or geometric scale ratio of the tank is:

$$\lambda_t = \frac{D_{t,1}}{D_{t,2}} = \left(\frac{v_{c,1}}{v_{c,2}}\right)^{1/3} = \left(\frac{5}{5{,}000}\right)^{1/3} = \frac{1}{10}$$

The diameter of the blades or turbine of the second tank (prototype) is:

$$D_{t,2} = 10 D_{t,1} = 10 \times 9.2 = 92\ cm$$

3) For a constant stirring speed N_t, if the diameter of the turbine D_t is multiplied by 10, the Reynolds number $\mathrm{Re} = \dfrac{\rho N_t D_t^2}{\mu}$ is multiplied by 100, and the regime remains turbulent.

The number of powers is constant ($N_{p,1} = N_{p,2}$), thus $\dfrac{p_{t,1}}{D_{t,1}^5} = \dfrac{p_{t,2}}{D_{t,2}^5}$, and therefore:

$$p_{t,2} = p_{t,1}\left(\frac{D_{t,2}}{D_{t,1}}\right)^5 = 5 \times \left(\frac{92}{9.2}\right)^5 = 500\ kW$$

This numerical result is prohibitive.

4) For a constant level of power per unit of volume, in a turbulent regime, $N_{t,1}^3 D_{t,1}^2 = N_{t,2}^3 D_{t,2}^2$ and therefore:

$$N_{t,2} = N_{t,1}\left(\frac{D_{t,1}}{D_{t,2}}\right)^{2/3} = 300 \times \left(\frac{1}{10}\right)^{2/3} = 64.63\ rpm$$

The Reynolds number of tank 2 expressed as a function of that of tank 1 gives:

$$Re_{t,2} = \frac{\rho N_{t,2} D_{t,2}}{\mu} = \frac{\rho (N_{t,1} / 10^{2/3})(10 D_{t,1})^2}{\mu} = \frac{10^2}{10^{2/3}} \frac{\rho N_{t,1} D_{t,1} D_{t,1}}{\mu}$$

$$Re_{t,2} = \frac{10^2}{10^{2/3}} \frac{\rho U_{t,1} D_{t,1}}{\mu} = 21.51 Re_{t,1}$$

As the regime still remains turbulent in the stirring tank, the power of the turbine is such that $p_t \propto N_t^3 D_t^5$.

The stirring power in tank 2 is:

$$p_{t,2} = p_{t,1}\left(\frac{N_{t,2}}{N_{t,1}}\right)^3\left(\frac{D_{t,2}}{D_{t,1}}\right)^5 = 5\times\left(\frac{64.63}{300}\right)^3\times\left(\frac{92}{9.2}\right)^5 = 5\ kW$$

With the points of the blade moving at a constant speed, in a turbulent regime, we obtain $N_{t,1}D_{t,1} = N_{t,2}D_{t,2}$.

The ratio of speeds is:

$$N_{t,2} = N_{t,1}\frac{D_{t,1}}{D_{t,2}} = 300\times\frac{1}{10} = 30\ rpm$$

The Reynolds number of tank 2 expressed as a function of that of tank 1 gives:

$$Re_{t,2} = \frac{\rho N_{t,2}D_{t,2}^2}{\mu} = \frac{\rho(N_{t,1}/10)(10D_{t,1})^2}{\mu} = \frac{10^2}{10}\frac{\rho N_{t,1}D_{t,1}D_{t,1}}{\mu}$$

$$Re_{t,2} = 10\frac{\rho U_{t,1}D_{t,1}}{\mu} = 10Re_{t,1}$$

The regime is fully turbulent, and therefore $p_t \propto N_t^3 D_t^5$.

The ratio of the powers of the two tanks allows us to write:

$$p_{t,2} = p_{t,1}\left(\frac{N_{t,2}}{N_{t,1}}\right)^3\left(\frac{D_{t,2}}{D_{t,1}}\right)^5 = 5\times\left(\frac{30}{300}\right)^3\times\left(\frac{10}{1}\right)^5 = 0.5\ kW$$

NOTE 3.19.– *Stirring is a unit operation that is required at many stages of the processes. Most of the stirring and mixing operations are carried out by means of a tilting stirrer that rotates around a shaft placed in a vessel, which is generally cylindrical in shape. The tanks are thus used from the storage of raw materials to that of finished products during operations of contact, reaction, and separation, which invariably occur during any process. The industries in which stirring is done are chemical, petrochemical, pharmaceutical, agribusiness, cosmetics, nuclear, metallurgy, water treatment and stationery.*

3.2.20. *Exercise 20: friction on a prototype probe*

Friction on a prototype probe must be estimated using data obtained using a model in a wind tunnel. The data and characteristics of the model and prototype are given in the table below:

Characteristics	Model sphere	Prototype sphere
Diameters	$D_m = 0.15\ m$	$D_p = 0.30\ m$
Friction force	$F_m = 25\ N$	$F_p = ?$
Fluid	Air	Water
Density	$\rho_m = 1.2\ kg/m^3$	$\rho_p = 10^3\ kg/m^3$
Dynamic viscosity	$v_m = 1.5 \times 10^{-5}\ m^2/s$	$v_m = 1.306 \times 10^{-6}\ m^2/s$
Speed of the fluid	$V_m = ?$	$V_p = 2.5\ m/s$

Table 3.25. *Data and characteristics of the similarity of the fluid flow around a prototype probe*

1) Determine the experimental velocity V_m required to obtain similarity between the two flows.

2) Determine the friction force F_p exerted by air on the prototype probe.

3.2.20.1. *Solutions*

1) Based on the dimensional analysis, it can be shown that the frictional force is given by the expression:

$$F = \rho \frac{V^2}{2} S f(Re)$$

with:

– $Re = \dfrac{VD}{\nu}$ is the Reynolds number;

– The function $f(Re)$ is the drag coefficient C_D which therefore varies on the basis of the number Re ($C_D = f(Re)$);

– $S = \dfrac{\pi D^2}{4}$ represents the apparent surface area of the sphere in the direction perpendicular to the flow;

– $\rho \dfrac{V^2}{2}$ the dynamic pressure.

The flows around the prototype sphere and the model sphere are in similarity, if the condition of Reynolds is checked, that is to say:

$$Re_m = Re_p \text{ or } \frac{V_m D_m}{\nu_m} = \frac{V_p D_p}{\nu_p}$$

And therefore:

$$V_m = \frac{\nu_m}{\nu_p}\frac{D_p}{D_m}V_p = \frac{1.5\times10^{-5}}{1.308\times10^{-6}}\times\frac{0.30}{0.15}\times2.5 = 57.38\ m/s$$

2) The condition of Reynolds similarity $Re_m = Re_p$ gives us:

$$f_m(Re_m) = f_p(Re_p) \text{ or } C_{D,m} = C_{D,p}$$

The friction force of the water on the prototype probe is:

$$F_p = F_m \frac{\rho_p}{\rho_m}\left(\frac{N_m}{N_p}\right)^3\left(\frac{D_m}{D_p}\right)^5 = 25\times\frac{10^3}{1.2}\times\left(\frac{2.5}{57.38}\right)^3\times\left(\frac{0.15}{0.30}\right)^5 = 158.18N$$

COMMENT 3.10.– *As an example of a measuring probe, consider the Pitot tube around which a fluid flows. This tube consists of two pressure taps:*

– a static pressure tap, where the measurement is carried out through the friction between the fluid and the tube wall;

– a total pressure tap, which gives the sum of the static pressure, pressure due to gravity (generally zero), and the dynamic pressure.

The flow velocity of the fluid is determined from the difference between the total pressure and the static pressure.

The Pitot tube depends on the direction and temperature of the fluid; it does not measure low flow velocities, its accuracy is limited, and it is sensitive to turbulent flows.

APPENDICES

Appendix 1

Some Dimensionless Numbers Used in Fluid Mechanics

Dimensionless number	Relationship	Type of relation and physical significance
Archimedes (buoyancy)	$Ar = \frac{gL_c^3\rho_f(\rho_f - \rho_c)}{\mu_f^2}$	Gravitational force × inertial force / viscous force
		Describes the movement of a body in a fluid, due to the difference in density between them
Capillary	$Ca = \frac{\mu_L V_c}{\sigma}$	Viscous force / surface tension force
		Used to describe liquid atomization
Capillarity	$Cap = \frac{\sigma k^{0.5}}{L_c V_f \mu_f}$	Capillary force / filtration force
		Used to characterize two-phase flows through porous materials
Ekman[1]	$Ek = \frac{\nu}{\omega L_c^2}$	Viscous force / Coriolis force
		Measurement of the attenuation of movement occurring in fluids due to rotation
Froude	$Fr = \frac{V_L}{\sqrt{gL_c}}$	Inertial force / gravitational force
		Characterizes the relative significance of the kinetic energy of the particles in a fluid relative to its gravitational potential energy

1 Vagn Walfrid Ekman, born on May 3, 1874, in Stockholm (Sweden) and who died on March 9, 1954, was an oceanographer.

Galileo[2]	$Ga = \sqrt{\frac{gL_c\rho_f(\rho_f - \rho_c)}{\mu_f^2}}$	Gravitational force / viscous force
		Used in fluid mechanics to quantify the ratio of the effects of buoyancy on viscous effects
Mach[3]	$Ma = \frac{V_f}{c}$	Fluid flow velocity / speed of sound
		Expresses the relationship of the local velocity of a fluid with respect to the speed of sound in the same fluid
Stokes	$St = \frac{1}{18}\frac{\rho_p d_p^2 V_f}{\mu_f L_c}$	Kinetic energy of the particle / energy dissipated by friction with the fluid
		Used in fluid dynamics to study the behavior of a particle in a fluid
Reynolds	$Re = \frac{\rho_f V_c L_c}{\mu_f}$	Inertial force / viscous force
		Describes a flow, particularly the nature of its type of regime (laminar, transient, turbulent)
Strouhal	$Sr = \frac{f L_c}{V_\infty}$	Inertial forces due to the local acceleration of the flow / inertial forces due to the convective acceleration
		Describes the oscillating flow mechanisms
Weber	$We = \frac{\rho_f L_c V_f}{\sigma}$	Inertia forces / tension surface forces
		Characterizes the flow of fluids at the interface of a multiphase system
Power	$Np = \frac{p}{\rho_f N^3 D^5}$	Drag forces / inertial forces
		Describes the conditions (energy consumption) of a mechanical mixture for a liquid

Table A1.1. *Dimensionless number used in fluid mechanics*

COMMENT ON TABLE A1.1.– c: speed of sound; D: diameter of stirrer; d_p: particle diameter; f: frequency of vortex shedding; g: gravitational acceleration; k: permeability of the medium; L_c: characteristic length;

2 Galileo Galilei, born in Pisa (Italy) in 1564, and who died on January 8, 1642, was a mathematician, geometrist, physicist and astronomer.

3 Ernst Mach, born on February 18, 1838, in Brno (Austria), and who lived until February 19, 1916, was a physicist and philosopher.

p: mechanical power; N: rotational speed; t: time; V_f: fluid velocity; V_c: characteristic velocity; V_L: liquid velocity; V_∞: freestream velocity; μ_f: dynamic viscosity of fluid; μ_L: dynamic viscosity of liquid; ν_f: kinematic viscosity of fluid; ρ_f: fluid density; ρ_p: particle diameter; ρ_c: body density; ρ_L: liquid density; ρ_p: particle density; σ: surface tension.

Appendix 2

Coefficients of Conversion to the International System or to the English System

English system	International system (SI)	SI Symbol	Conversion of the English system to SI by multiplying by:	Conversion of the SI to the English system by multiplying by:
Length				
Inch	Centimeter	cm	2.54	0.3937
Foot	Meter	m	0.3048	3.281
Mile	Kilometer	km	1.6039	0.6214
Surface area				
Square inch	Square centimeter	cm^2	6.452	0.1550
Square foot	Square meter	m^2	0.09290	10.76
Acre	Hectare	ha	0.4047	2.471
Volume				
Cubic inch	Cubic centimeter	cm^3	16.387	0.06102
Cubic foot	Cubic meter	m^3	0.02832	35.32
Gallon	Cubic meter	m^3	0.003785	264.2
Gallon	Liter	l	3.875	0.2642

Mass				
Pound mass	Kilogram	kg	0.4536	2.205
Slug	Kilogram	kg	14.59	0.06854
Density				
Slug/cubic foot	Kg/cubic meter	Kg/m^3	515.4	1.94×10^{-3}
Force				
Pound	Newton	N	4.448	0.2248
Kip (1,000 lb)	Newton	N	4448	-
Work, energy, heat				
Foot-pound	Joule	J	1.356	0.7376
BTU	Kilojoule	kJ	1.054	0.9479
BTU	Kilowatt-hour	kWh	0.000293	3413
Therm	Kilowatt-hour	kWh	29.3	0.03413
Power, heat flow				
Horsepower	Kilowatt	kW	0.7457	1.341
Foot-pound/sec	Watt	W	1.356	0.7376
BTU/hour	Watt	W	0.2929	3.414
Pressure				
Pound/square inch	Kilopascal	kPa	6.895	0.1450
Pound/square foot	Kilopascal	kPa	0.04788	20.89
Feet of H_2O	Kilopascal	kPa	2.983	0.3352
Feet of Hg	Kilopascal	kPa	3.374	0.2964
Temperature				
Fahrenheit	Celsius	°C	5/9(°F - 32)	5/9 x °C + 32
Fahrenheit	Kelvin	K	5/9(°F + 460)	5/9 x K - 460

English system	International system (SI)	SI Symbol	Conversion of the English system to SI by multiplying by:	Conversion of the SI to the English system by multiplying by:
Speed				
Foot/second	Meter/second	m/s	0.3048	3.281
Mile/hour	Meter/second	m/s	0.4470	2.237
Mile/hour	Kilometer/hour	km/h	1.609	0.6215
Acceleration				
Foot/second squared	Meter/second squared	m/s^2	0.3048	3.281
Torque				
Pound-foot	Newton-meter	N.m	1.356	0.7376
Pound-inch	Newton-meter	N.m	0.1130	8.85
Dynamic viscosity, kinematic viscosity				
Pound-sec/ Square foot	Newton-second/ square meter	$N.s/m^2$	47.88	0.02089
Square foot/ Second	Square meter/second	m^2/s	0.09290	10.76
Volumetric flow				
Cubic-foot/second	Cubic meter/second	m^3/s	0.02832	35.32
Cubic-foot/second	Liter/second	l/s	28.32	0.03532

Table A2.1. *Coefficients of conversion to the International system (SI) or to the English system*

EXAMPLE.–

– $1\ mile = 1.6093 \times 1\ km$; $1\ km = 0.6214\ mile$: follow the same reasoning for other units.

– $1\ horse\ power = 0.7457 \times 1\ kW$; $1\ kW = 1.341 \times 1\ horse\ power$: follow the same reasoning for other units.

References

[BOU 98] BOUTTES J., *Mécanique des fluides*, Ellipse, Paris, 1998.

[CAN 95] CANDEL S., *Problèmes de mécanique des fluides*, Dunod, Paris, 1995.

[CAR 98] CARLIER M., *Hydraulique générale et appliquée*, Eyrolles, Paris, 1998.

[CHA 10] CHASSAING P., *Mécanique des fluides: éléments d'un premier parcours*, 3rd edition, Cépaduès, Toulouse, 2010.

[COM 79] COMOLET R., BONNIN J., *Mécanique expérimentale des fluides*, volume 1, 3rd edition, Masson, Paris, 1979.

[COM 81] COMOLET R., BONNIN J., *Mécanique expérimentale des fluides*, volume 3, 3rd edition, Masson, Paris, 1981.

[COM 06] COMOLET R., *Mécanique expérimentale des fluides*, volume 2, 4th edition, Dunod, Paris, 2006.

[CRA 01] CRASSARD J.J., La Pompe centrifuge dans tous ses états, chauffage, sanitaire, climatisation, alimentation, surpression, relevage, EDIPA, Paris, 2001.

[DAR 14] DARROZES J.S., MONAVON A., *Analyse phénoménologique des écoulements – Comment traiter un problème de mécanique des fluides avant de résoudre les équations*, Presses polytechniques et universitaires romandes, Lausanne, 2014.

[FAU 08] FAURE T., *Dynamique des fluides appliquée – Application à l'aérodynamique*, Dunod, Paris, 2008.

[HOU 75] HOUPEURT A., *Éléments de mécanique des fluides dans les milieux poreux*, Technip, Paris, 1975.

[HUG 75] HUG M., *Mécanique des fluides appliquée aux problèmes d'aménagement et d'énergétique*, 1st part, volume 1, Eyrolles, Paris, 1975.

[JOH 17] JOHN M.C., YUNUS A.C., *Mécanique des fluides: fondements et applications*, Deboeck, Paris, 2017.

[LAN 98] LANDAU L., LIFCHITZ E., *Physique théorique: mécanique des fluides*, 3rd edition, Eyrolles, Paris, 1998.

[LEN 99] LENCASTRE A., *Hydraulique générale*, Eyrolles, Paris, 1999.

[LUM 96] LUMBROSO H., *Mécanique des fluides*, Dunod, Paris, 1996.

[MOR 92] MOREL M.A., ABORDE J.P., *Exercices de mécanique des fluides*, volume 1, Eyrolles, Paris, 1992.

[MOR 94] MOREL M.A., ABORDE J.P., *Exercices de mécanique des fluides*, volume 2, Eyrolles, Paris, 1994.

[OUZ 98] OUZIAUX R., PERRIER J., *Mécanique des fluides appliquée*, 3rd edition, Dunod, Paris, 1998.

[PAS 15] PASCAL B., RICHARD M., *Mécanique des fluides en 20 fiches*, Dunod, Paris, 2015.

[POT 08] POTTER M., WIGGERT D.C., *Fluid mechanics*, McGraw-Hill, New York, 2008.

[PLU 10] PLUVIOSE M., *Machines à fluides – Principes et fonctionnement*, Ellipses, Paris, 2010.

[ROB 96] ROBERSON J.A., ACOSTA A.J., *Engineering fluid mechanics*, 6th edition, Houghton-Mifflin, Boston, 1996.

[SAB 01] SABERSKY R.H., ACOSTA A.J., HAUPTAM E.G., *Fluid Flow*, 5th edition, Macmillan, New York, 2001.

[SAK 17] SAKIT M., JEAN-LUC B., *Mécanique des fluides*, 3rd edition, Dunod, Paris, 2017.

[STR 85] STREETER V.L., WYLIE E.B., *Fluid Mechanics*, 8th edition, McGraw-Hill, New York, 1985.

Index

U

V, W

Other titles from

iSTE

in

Fluid Mechanics

2019

CALTAGIRONE Jean-Paul
Discrete Mechanics: Concepts and Applications

2018

ABABOU Rachid
Capillary Flows in Heterogeneous and Random Porous Media 1

2017

TARDU Sedat
Wall Turbulence Control

2015

CALTAGIRONE Jean-Paul
Discrete Mechanics

FOHR Jean-Paul
Heat and Moisture Transfer between Human Body and Environment

2014

PRUD'HOMME Roger
Flows and Chemical Reactions in Heterogeneous Mixtures
Flows and Chemical Reactions in an Electromagnetic Field

TARDU Sedat
Transport and Coherent Structures in Wall Turbulence

2013

BORGHI Roland, ANSELMET Fabien
Turbulent Multiphase Flows with Heat and Mass Transfer

DÉLERY Jean
Three-dimensional Separated Flow Topology: Critical Points, Separation Lines and Vortical Structures

PRUD'HOMME Roger
Flows and Chemical Reactions in Homogeneous Mixtures

2012

BOUTIER Alain
Laser Velocimetry in Fluid Mechanics

BOUTIER Alain
Laser Metrology in Fluid Mechanics: Granulometry, Temperature and Concentration Measurements

PRUD'HOMME Roger
Flows and Chemical Reactions

2011

MORY Mathieu
Fluid Mechanics for Chemical Engineering

TARDU Sedat
Statistical Approach to Wall Turbulence

2010

COLIN Stéphane
Microfluidics

DÉLERY Jean
Handbook of Compressible Aerodynamics

GUINOT Vincent
Waves Propagation in Fluids – 2nd edition

SOULI Mhamed, BENSON J. David
Arbitrary Lagrangian Eulerian and Fluid-structure Interaction

2009

FAVRE-MARINET Michel, TARDU Sedat
Convective Heat Transfer

2008

LALAUZE René
Physical Chemistry of Solid-Gas Interfaces

PEUBE Jean-Louis
Fundamentals of Fluid Mechanics and Transport Phenomena

SCHIESTEL Roland
Modeling and Simulation of Turbulent Flows

www.ingramcontent.com/pod-product-compliance
Lightning Source LLC
LaVergne TN
LVHW011801200926
848484LV00003B/747